Wastewater Residuals Stabilization

Manual of Practice FD-9

Prepared by the Task Force on Wastewater Residuals Stabilization

Jane B. Forste, *Chair*
Max L. Anderson
A. Ron Appleton, Jr.
Philip N. Baldwin
Robert K. Bastian
Robert D. Blythe
Robert J. Bolton
Douglas R. Borgatti
David L. Bowers
Edward M. Boyle
Jeffrey Burnham
Timothy Carey
John Chack
Sidney E. Clark
Craig S. Coker
Alan B. Cooper
Tracy Cork
John F. Donovan
Donald D. Dorsey
Daniel D. Durfee
Brad J. Einfeld

Arthur H. Fagerstrom
Joseph B. Farrell
Gerald W. Foess
Thomas M. Getting
James R. Glaser
Padma S. Golla
Mark Gould
Brian W. Hemphill
Joseph A. Husband
Allan Jacobs
Stephen Kellogg
Victor Khoury
Agamemnon Koutsospyros
Stephen M. Lacy
Douglas A. Logan
Christopher Mahoney
Brian F. McNamara
Brooks W. Newbry
Miryoussef Norouzian
David W. Oerke
Dan L. Pecha

Robert S. Reimers
Clayton M. Richardson
Glenn A. Richter
Francis Riddle
Julian Sandino
Dennis E. Sarpen
Paul D. Saurer
Perry L. Schafer
Harold E. Schmidt, Jr.
Kenneth Schnaars
Ronald B. Sieger
Michael Spence
Srinivasan Sundaramoorthy
Rao Y. Surampalli
Mohan V. Thampi
K. Richard Tsang
Thomas E. Vik
T. Viraraghavan
Milind V. Wable
Bethanne B. Williams
William O. Wurtz

Under the Direction of the **Facilities Development Subcommittee of the Technical Practice Committee**

1995

Water Environment Federation
601 Wythe Street
Alexandria, Virginia 22314–1994

Library of Congress Cataloging-in-Publication Data
Wastewater residuals stabilization / prepared by the Task Force on Waste-
    water Residuals Stabilization.
        p. cm.—(Manual of practice; FD-9)
    Includes bibliographical references and index.
    ISBN 1-881369-94-3
1. Sewage sludge digestion.   I. Water Environment Federation. Task Force
on Wastewater Residuals Stabilization.   II. Series: Manual of practice. FD;
no. 9.
TD769.S57 1995
628.3—dc20   95-14028
                                                                   CIP

Manuals of Practice for Water Pollution Control

The Water Environment Federation (WEF) Technical Practice Committee (formerly the Committee on Sewage and Industrial Wastes Practice of the Federation of Sewage and Industrial Wastes Associations) was created by the Federation Board of Control on October 11, 1941. The primary function of the committee is to originate and produce, through appropriate subcommittees, special publications dealing with technical aspects of the broad interests of the Federation. These manuals are intended to provide background information through a review of technical practices and detailed procedures that research and experience have shown to be functional and practical.

IMPORTANT NOTICE

The contents of this publication are for general information only and are not intended to be a standard of the Water Environment Federation (WEF).

No reference made in this publication to any specific method, product, process, or service constitutes or implies an endorsement, recommendation, or warranty thereof by the Federation.

The Federation makes no representation or warranty of any kind, whether expressed or implied, concerning the accuracy, product, or process discussed in this publication and assumes no liability.

Anyone using this information assumes all liability arising from such use, including but not limited to infringement of any patent or patents.

Water Environment Federation Technical Practice Committee Control Group

L.J. Glueckstein, *Chair*
T. Popowchak, *Vice-Chair*

P.T. Karney
T.L. Krause
J. Semon
R. Zimmer

Authorized for Publication by the Board of Control
Water Environment Federation

Quincalee Brown, *Executive Director*

Preface

Stabilization of the residual solids removed during treatment is key to the reliable performance of any plant. Dealing with these solids—removing, thickening, stabilizing, dewatering, and disposing them—is often a technical and operational challenge. It is imperative that the designer recognize the significance of the solids-related portion of the plant and be diligent in its design.

This manual serves as a primer for the design of the more commonly used stabilization processes. No attempt is made to present design standards nor are the less widely used and newer stabilization technologies presented. Process operation and maintenance are included to the extent that they affect design, and Water Environment Federation (WEF) Manual of Practice No. 11, *Operation of Municipal Wastewater Treatment Plants,* should be consulted for these aspects.

This manual also serves as an important segue for the changes in terminology currently taking place in the solids management industry as a result of the negative public reaction often experienced when the term "sludge" is used. In 1992, to promote public acceptance of reuse projects, the Federation endorsed the term "biosolids" to define the primarily organic solids product of municipal wastewater treatment that can be beneficially used. In describing the stabilization processes in this manual, "sludge" is used only to describe the unstabilized biomass, and "biosolids" is used to connote the beneficially usable product.

This publication was developed under the direction of Jane B. Forste, *Chair.* Principal authors of the manual are

A. Ron Appleton, Jr.	John F. Donovan	Julian Sandino
Philip N. Baldwin	Daniel D. Durfee	Kenneth Schnaars
Robert D. Blythe	Mark Gould	Thomas E. Vik
David L. Bowers	Stephen Kellogg	Milind V. Wable
Edward M. Boyle	David W. Oerke	Bethanne B. Williams
Tracy Cork	Robert S. Reimers	

Additional information and review were provided by

Dan Buhrmaster	Rob Christy	Steve Rogowski
Jeff Burnham	Bob Fergen	Lori Stone
Paul Christy	Terry Logan	

The efforts of the authors and reviewers were supported by the following organizations:

Arvin North American Auto, Columbus, Ohio
Bio Gro Systems, Inc., Annapolis, Maryland
Bird Environmental Technologies, Trumbull, Connecticut

Black & Veatch, Inc., Aurora, Colorado; Kansas City, Missouri
Brown & Caldwell, Inc., Sacramento, California; Denver, Colorado
Camp Dresser & McKee, Inc., Cambridge, Massachusetts; Raleigh, North
  Carolina
CH2M Hill, Inc., Deerfield Beach, Florida; Corvallis, Oregon
Consoer, Townsend, & Assoc., Inc., Nashville, Tennessee
CSI, Maitland, Florida
Dyer, Riddle, Mills & Precourt, Tampa, Florida
Engineering–Science, Inc., Fairfax, Virginia
F B Leopold Company, Zelienople, Pennsylvania
FHC Engineers, Tulsa, Oklahoma
Hampton Roads Sanitation District, Virginia Beach, Virginia
Hartman & Associates, Inc., Orlando, Florida
Hazen & Sawyer, Inc., New York, New York
HDR Engineering, Inc., Denver, Colorado; Lake Oswego, Oregon; Austin,
  Texas
Hydropress N-Viro Services, Inc., Hatfield, Massachusetts
Jacobs Environmental, Inc., Piscataway, New Jersey
James R. Holly & Assoc., Inc., York, Pennsylvania
Lewiston–Auburn WPCA, Lewiston, Maine
Malcolm Pirnie, Inc., Phoenix, Arizona; White Plains, New York
McMahon Associates, Inc., Madison, Wisconsin
Metcalf & Eddy, Inc., New York, New York
Montgomery Watson, Walnut Creek, California; Herndon, Virginia
Philip Giantris & Associates, North Andover, Massachusetts
PORI International, Inc., Baltimore, Maryland
Professional Services Group, Erdenheim, Pennsylvania
PWT Waste Solutions, Houston, Texas
Radian Corporation, Denver, Colorado
Tulane University, New Orleans, Louisiana
U.S. Environmental Protection Agency, Washington, D.C.
University of New Haven, West Haven, Connecticut
University of Regina, Regina, Saskatchewan, Canada
University of Wisconsin–Platteville, Wisconsin
Village of Minerva Park, Columbus, Ohio

Federation technical staff review was provided by Berinda J. Ross. Technical editorial assistance was provided by Jodi L. Bergeman.

Contents

List of Tables

List of Figures

Chapter 1
Stabilization Overview

This manual emphasizes stabilization theory, research results, and process design for stabilization processes such as anaerobic digestion, aerobic digestion, composting, and lime stabilization. New processes such as autothermal aerobic digestion, alkaline stabilization, heat drying, and other emerging technologies are also described. Some technologies such as irradiation and super chlorination are not discussed in this publication. This manual is intended to complement Manual of Practice No. 8, *Design of Municipal Wastewater Treatment Plants* (1992), and Manual of Practice No. 11, *Operation of Municipal Wastewater Treatment Plants* (1990), published by the Water Environment Federation.

The Water Environment Federation defines biosolids as the primarily organic solid product of municipal wastewater treatment that can be beneficially used. The stabilization processes discussed in this manual are those typically used in municipal operations to produce biosolids. It should be noted, however, that the biosolids produced, in some cases, may not meet current federal or state standards for land application. Further treatment may be needed to satisfy regulatory criteria.

In this publication, the term "biosolids" is reserved exclusively to refer to the stabilized product. Unstabilized materials are referred to by several terms, including "sludge," "solids," "cake," and "residuals."

*D*ESIGN APPROACH TO STABILIZATION PROCESSES

Stabilization processes are key to the reliable performance of any wastewater treatment plant (WWTP). Historically, design engineers have concentrated on cleaning up the water and considered solids management as a secondary issue. In reality, removing solids from the liquid, thickening, stabilizing, dewatering them, and using or disposing of biosolids is often a more difficult challenge, both technically and operationally. Further, improper solids management has the potential to create damage to the environment sometimes in the magnitude equal to the discharge of untreated wastewater. In addition, solids-related facilities can account for nearly 50% of the total treatment cost. Design engineers should recognize the significance of the solids-related portion of the WWTP and be as diligent in its design as for the remainder of the processes.

When designing a stabilization process, it is important that the designer consider the quantity of solids to be treated. The designer should not only design the solids-handling portion of the WWTP for average production rates, but also the monthly maximum. The designer should evaluate various stabilization processes that can be easily integrated into the WWTP scheme. Reviewing various aspects of the process, such as the effects from the stabilization process sidestreams, whether the facility is striving to achieve a high-quality product for beneficial use, safety, ease of operation, ancillary equipment needs, and other similar parameters are important to ensure a well-designed system. The designer must use a system approach that investigates both the economic and noneconomic considerations of the process being designed or evaluated.

*Q*UANTITIES TO BE TREATED

Whenever possible, design engineers should compare historical data from a specific WWTP site against reported data from other plants that have forecasted quantities of residual solids to be treated. Industrial discharges can have a significant effect on quality as well as quantity. Developing a detailed solids balance and taking into account the various factors that affect the quantity are also important considerations that must be investigated. Figure 1.1 and Table 1.1 give an example of a solids flow sheet and mass balance, a convenient method for presenting calculations.

SYSTEMS APPROACH

Stabilization is only one process in the system. Thickening, stabilization, and dewatering processes should be compatible with each other, and the overall system design should be integrated. For example, adequate thickening before stabilization may allow the stabilization facilities to be downsized.

OBJECTIVES

The engineer should know the solids disposal methods to be used. Stabilized biosolids characteristics that are appropriate for this disposal (or beneficial use) option should be determined. Determine whether capital cost, operating cost, or life cycle cost is most important. Check with the regulating agencies to determine whether federal, state, or local regulations may directly or indirectly affect the process selection, design criteria, redundancy requirements, or disposal restrictions.

Regulations governing the disposal of wastewater residuals or use of biosolids on land were promulgated by the U.S. Environmental Protection Agency (U.S. EPA) on February 19, 1993, with the publication of the Final Rule *Standards for the Use or Disposal of Sewage Sludge.* These regulations are published in the Code of Federal Regulations (CFR), Title 40, Parts 257, 258, 403, and 503. Basically, Part 257 establishes rules for residual solids generated at industrial facilities and domestic septage combined with commercial industrial septage (U.S. EPA, 1993). Part 258 focuses on disposal in a municipal solid waste landfill. Part 503 defines the standards for the use and disposal of biosolids for land application, surface disposal, and incineration. The

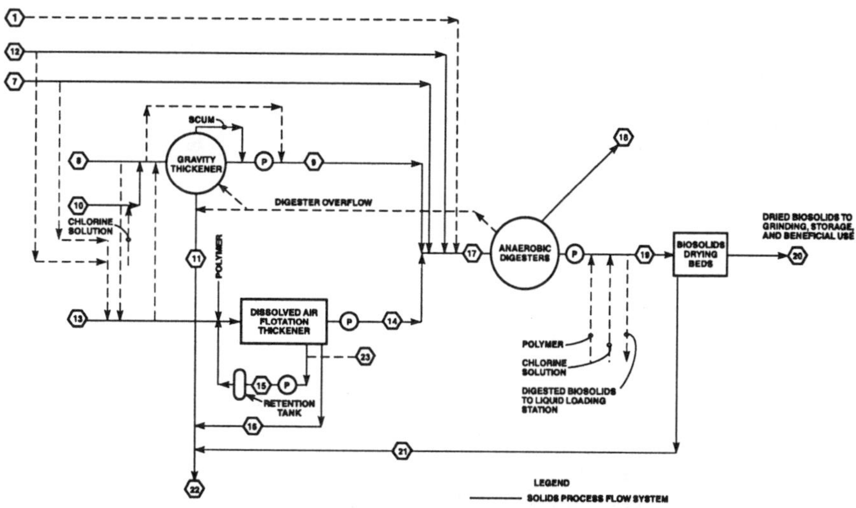

Figure 1.1   **Example solids process diagram.**

Table 1.1   Example liquid and solids balance.

Stream identification	Dry solids, %		Liquid flow rate, gpd			Biochemical oxygen demand, lb/d			Total suspended solids, lb/d		
	Average	Range	Minimum	Average	Peak	Annual average	Peak month average	Peak week average	Annual average	Peak month average	Peak week average
Septage	3.6	0.5 – 10	0	1 105	7 000	200	240	300	350	430	500
Inplant wastewater	0.02	0.01 – 0.04	0	250	700	10	10	10	10	10	10
Screenings	80	70 – 90	23[a]	64[a]	149[a]	—	—	—	3 940[b]	8 320[b]	8 940[b]
Grit slurry	—	—	1 940[c]	16 200[c]	179 000[c]	—	—	—	—	—	—
Grit	90	90 – 95	15[a]	64[a]	166	—	—	—	9 300[b]	22 600[b]	24 400[b]
Grit wash water	—	—	1 830	14 500	177 000	—	—	—	—	—	—
Primary scum	0.05	0.01 – 0.5	4 320[c]	7 210[c]	18 800[c]	—	—	—	—	—	—
Primary sludge	3.5	2 – 6	32 000	55 900	150 000	7 115	8 215	10 270	16 316	18 715	24 905
Thickened primary sludge	6	6 – 10	18 000	23 200	48 000	6 006	7 306	9 245	10 500	17 780	23 880
Sweetener water	—	—	427 000	621 000	738 000	—	—	—	—	—	—
Gravity thickener overflow	0.02	0.0005 – 0.10	529 000	554 000	752 000	710	820	1 025	815	935	1 245
Secondary scum	0.06	0.01 – 0.5	6 240[c]	8 730[c]	19 200[c]	—	—	—	—	—	—
Waste-activated sludge	0.7	0.5 – 1.5	96 000	210 000	475 000	—	—	—	12 270	15 995	19 940
Thickened activated sludge	3.5	2.5 – 7	18 000	35 700	90 000	—	—	—	10 430	15 070	18 950
Flotation thickener recycle	0.01	0 – 0.06	345 000	515 000	1 340 000	—	—	—	—	—	—
Flotation thickener underflow	0.13	0.01 – 0.2	70 000	174 000	440 000	1 360	695	745	1 840	795	900
Blended sludge	5.3	4 – 8	36 000	58 000	138 000	—	—	—	25 930	32 960	42 510
Digested gas	—	—	110 000[a]	152 000[a]	640 000[a]	—	—	—	—	—	—

Table 1.1   Example liquid and solids balance (continued).

Stream identification	Dry solids, %		Liquid flow rate, gpd			Biochemical oxygen demand, lb/d			Total suspended solids, lb/d		
	Average	Range	Minimum	Average	Peak	Annual average	Peak month average	Peak week average	Annual average	Peak month average	Peak week average
Digested biosolids	3.4	2.5 – 6	36 000[d]	68 900[d]	138 000[d]	—	—	—	16 676	21 380	28 705
Dewatered sludge	50	36 – 65	400[a]	630[a]	1 300[a]	—	—	—	16 510	21 135	29 420
Drying bed underflow	0.04	0.01 – 0.1	33 000	55 000	128 000	35	110	145	165	215	206
Thickener/drying bed return flows	0.04	0.01 – 0.1	630 000	783 000	1 310 000	2 175	1 525	1 915	2 620	1 945	2 520
Auxiliary water supply	—	—	0	0	1 440 000	—	—	—	—	—	—

Note: gpd $\times$ 4.383 $\times$ 10^{-5} = L/s; lb/d $\times$ 0.453 6 = kg/d.
[a] Expressed in cu ft/d; cu ft/d $\times$ 28.32 = cu L/d.
[b] Wet weight.
[c] Not continuous.
[d] Assumes no supernating.

previous (40 CFR Part 257) regulations divided processes into those that significantly reduce pathogens (PSRP) and those that further reduce pathogens (PFRP).

The Final Rule describes the frequency of monitoring, land application, incineration, reporting, pathogen reduction, vector attraction, recordkeeping, and other criteria for the use or disposal of biosolids. The 40 CFR Part 503 regulations divide the quality of the biosolids into two categories, referred to as Class A and Class B. Essentially, the Class A biosolids are required to meet certain criteria to ensure they are safe for use by nurseries, golf courses, or the general public. Class B biosolids have lesser treatment requirements than Class A, and typically are distributed onto agricultural land or into a landfill.

A Plain English Guide to the EPA Part 503 Biosolids Rule (U.S. EPA, 1994) provides a helpful tool for interpretation and implementation of the rule. Appendix A of this manual includes a summary of the processes used to produce Class A and B biosolids and of the pathogen and vector attraction reduction requirements for both classes.

DESIGN

Typically, the differences between a successful WWTP and an unsuccessful WWTP are details overlooked in design, including flexibility, instrumentation and control devices, accessibility, and ease of maintenance. Several techniques can be used to provide an operations focus to the design, such as involving an operator in conceptual design, drafting the operations manual before finishing the design, and conducting a value engineering analysis of the conceptual design. These techniques force the designer to think through the operation of the facility while there is still time to make design improvements to enhance operability.

In addition to the normal safety hazards associated with mechanical and electrical equipment, explosive and toxic gases are generated in anaerobic digestion. *Safety and Health in Wastewater Systems* discusses plant safety in some detail (WEF, 1994). Individual chapters in the manual discuss specific safety concerns associated with each process. During design, consider a formal designer and owner review or workshop dedicated to providing a safe facility. In addition, many states employ occupational safety engineers who may be willing to provide a review of the facility design free of charge. The designer must consider all federal, state, and local codes in the design. The designer should also contact the local fire marshal and follow all fire codes, such as the National Fire Protection Association Code (NFPA, 1992).

*C*OMPARISON OF PROCESSES

Anaerobic digestion is perhaps the most widely used of the processes discussed in this manual. It produces relatively stable biosolids at a moderate cost and, as an added benefit, produces methane gas. This gas can be used to heat the digester, facility, or both and in larger WWTPs may even be used for cogeneration (for example, a gas-engine-driven generator with jacket water heat recovery). Some of the major disadvantages of this process include high initial cost; a significant amount of mechanical equipment (particularly where digester gas is used); a strong sidestream (supernatant); the necessity for heat input to maintain the desired temperature; and the tendency toward process upset as a result of poor mixing, overloading, lack of temperature control, and heavy metals or other toxic agents in the feed.

Aerobic digestion is typically found in smaller WWTPs, of less than 200 L/s (5 mgd). It is a power-intensive process (because of power requirements for oxygen transfer) compared with anaerobic digestion, but it is typically also less expensive at the outset. Aerobic digestion typically is less complex operationally and is sometimes not even a separate process. In many extended aeration facilities, such as oxidation ditches, the solids retention time is maintained long enough to provide digestion by endogenous respiration in the aeration system. However, because of new regulations a number of states are not allowing treatment facilities to use the volatile solids reduction in the aeration tanks as part of the 38% volatile solids reduction required in the stabilization process.

Autothermal thermophilic aerobic digestion (ATAD) is a stabilization process that operates at thermophilic temperatures of 40 to 80°C using aerobic microorganisms (U.S. EPA, 1990). Initial studies began in the 1960s, and testing in the 1970s; however, it has only been recently that successful facilities have been installed. Refinements and lessons learned from the first facilities have helped contribute to the success of the recently constructed ATAD facilities.

Composting typically is used where biosolids use as a soil amendment/conditioner has been determined to be appropriate. Feed to the composting process can be either raw or digested. The final compost mixture should have a minimum solids content of 40%. A bulking agent is added to attain this solids level, the desired structural properties of the mass, and to promote adequate air circulation. Composting typically is a labor-intensive process because of the addition of the bulking agent, turning of the windrows (for windrow composting), and recovery of the bulking agent. The potential for odors exists, especially at poorly designed and operated sites. Additionally, there is a possibility of pathogen spread through dust.

Lime-stabilized biosolids typically are used to achieve a Class B product, but in some instances they are used as a soil amendment/conditioner and used to meet a Class A status. The process is inexpensive when achieving a

---

Class B product and is simple to operate compared with digestion and composting. Some of the principal disadvantages are that the biosolids can become unstable if the pH drops after treatment and regrowth of biological organisms occurs, the added lime increases the mass of solids to be used or disposed, and the lime is expensive.

In recent years, a number of advanced alkaline stabilization technologies have emerged in the wastewater field that use chemical additives in addition to or instead of lime. These processes use chemicals such as cement kiln dust, lime kiln dust, Portland cement, fly ash, and other types of additives to achieve PFRP criteria. Alkaline stabilization processes meet Class A requirements for beneficial use. As with all stabilization technologies, alkaline stabilization processes have their advantages and disadvantages. For example, an advantage to these processes is that they produce a rich soillike product with substantially reduced pathogens. One disadvantage to this type of process is that it increases the end product mass rather than reducing the mass.

Thermal drying is a technology gaining acceptance throughout the wastewater industry. Thermal dryers currently used in the U.S. are direct dryers, indirect dryers, and multiple effect evaporation dryers. Feed solids concentrations to the dryers range from 20 to 25%; outlet concentrations vary from 40 to 98%. When dried to 40%, the solids typically are further processed using composting, advanced alkaline stabilization, or incineration. At a concentration of 98%, however, the solids meet the 40 CFR Part 503 requirements for Class A biosolids and can be beneficially used for land application or distribution and marketing.

Dryers operate at high temperatures to drive off the moisture and use either steam, gas, fuel oil, or infrared lamps as their heating source. The offgases from the dryers are treated in wet scrubber odor control systems before being released into the atmosphere.

The release of the 40 CFR Part 503 regulations and the interest generated by the public's concern for the environment has increased research into new technologies for beneficial use of biosolids. The enactment of the 40 CFR Part 503 regulations has prompted many engineers and municipalities to investigate or design effective beneficial use systems.

Table 1.2 summarizes many of the advantages and disadvantages of the principal stabilization processes.

Table 1.2   Comparison of stabilization processes.

Process	Advantages	Disadvantages
Anaerobic digestion	Good volatile suspended solids destruction (40 to 60%) Net operational costs can be low if gas (methane) is used Broad applicability Biosolids suitable for agricultural use Good pathogen inactivation Reduces total sludge mass Low net energy requirements	Requires skilled operators May experience foaming Methane formers are slow growing, hence "acid digester" sometimes occurs Recovers slowly from upset Supernatant strong in carbonaceous oxygen demand, biochemical oxygen demand, suspended solids, and ammonia Cleaning is difficult (scum and grit) Can generate nuisance odors resulting from anaerobic nature of process High initial cost Potential for struvite (mineral deposit) Safety issues concerned with flammable gas
Aerobic digestion	Low initial cost, particularly for small plants Supernatant less objectionable than anaerobic Simple operational control Broad applicability If properly designed, does not generate nuisance odors Reduces total sludge mass	High energy costs Generally lower volatile suspended solids destruction than anaerobic Reduced pH and alkalinity Potential for pathogen spread through aerosol drift Biosolids typically are difficult to dewater by mechanical means Cold temperatures adversely affect performance May experience foaming
Autothermal thermophilic aerobic digestion	Reduced hydraulic retention compared with conventional aerobic digestion Volume reduction Excess heat can be used for building heat Pasteurization of the sludge, pathogen reduction	High energy costs Potential of foaming Requires skilled operators Potential for odors

Table 1.2   Comparison of stabilization processes (continued).

Process	Advantages	Disadvantages
Composting	High quality, potentially saleable end product suitable for agricultural use Can be combined with other processes Low initial cost (static pile and windrow)	Requires 18 to 30% dewatered solids Requires bulking agent Requires either forced air (power) or turning (labor) Potential for pathogen spread through dust High operational cost; can be power-, labor-, or chemical-intensive, or all three May require significant land area Requires carbon source Potential for odors
Lime stabilization	Low capital cost Easy operation Good as interim or emergency stabilization method	Biosolids not always appropriate for land application Chemical-intensive Overall cost very site-specific Volume of biosolids to be disposed of is increased pH drop after treatment can lead to odors and biological growth
Advanced alkaline stabilization	Produces a high quality Class A end product Can be started quickly Excellent pathogen reduction	Operator intensive Chemical intensive Potential for odors Volume of biosolids to be disposed of is increased May require significant land area
Sludge dryers	Substantially reduces volume Can be combined with other processes Produces a Class A product Not a biological process so it can be started quickly Retains nutrients	Some dryers could be labor-intensive Produces an offgas that must be treated

R*EFERENCES*

National Fire Protection Association (1992) *Fire Protection in Wastewater Treatment and Collection Facilities.* Vol. 10, Sect. 820, Quincy, Mass.

U.S. Environmental Protection Agency (1990) *Autothermal Thermophilic Aerobic Digestion of Municipal Wastewater Sludge.*

U.S. Environmental Protection Agency (1993) *Standards for the Use or Disposal of Sewage Sludge. Fed. Regist.,* **58,** 32, 40 CFR Parts 257, 258, 403, and 503.

U.S. Environmental Protection Agency (1994) *A Plain English Guide to the EPA Part 503 Biosolids Rule.* EPA-832/R-93-003, Off. of Wastewater Manage., Washington, D.C.

Water Environment Federation (1990) *Operation of Municipal Wastewater Treatment Plants.* Manual of Practice No. 11, Alexandria, Va.

Water Environment Federation (1992) *Design of Municipal Wastewater Treatment Plants.* Manual of Practice No. 8, Alexandria, Va.; Am. Soc. Civ. Eng., Manual and Report on Engineering Practice No. 76, New York, N.Y.

Water Environment Federation (1994) *Safety and Health in Wastewater Systems.* Manual of Practice No. 1, Alexandria, Va.

Chapter 2
Aerobic Digestion

Stabilization during the aerobic digestion process occurs from the destruction of degradable organic components and the reduction of pathogen organisms by aerobic, biological mechanisms. The aerobic digestion process is a suspended-growth biological treatment process and is based on biological theories similar to those of the extended aeration modification of the activated-sludge process.

The objectives of the aerobic digestion process, which can be compared with those of the anaerobic digestion process, include production of a stable product by oxidizing organisms and other biodegradable organics, reduction of mass and volume, reduction of pathogen organisms, and conditioning for further processing. Advantages of the aerobic process compared with anaerobic digestion are the production of an inoffensive, biologically stable end product; lower capital costs; more simple operational control with reductions in volatile solids concentrations slightly less than those achieved in the anaerobic digestion process; safer operation with no potential for gas explosion and less potential for odor problems; and discharge of a supernatant with a

5-day biochemical oxygen demand test (BOD_5) concentration typically less than that found in the anaerobic process.

The primary disadvantage typically attributed to the aerobic digestion process is the higher power cost associated with oxygen transfer. Recent developments in the aerobic digestion process, such as highly efficient oxygen transfer equipment and research into operation at elevated temperatures, may reduce this concern. Other disadvantages cited include the reduced efficiency of the process during cold weather, the inability to produce a useful byproduct such as methane gas from the anaerobic process, and the mixed results achieved during mechanical dewatering of aerobically digested biosolids.

A review of the current literature regarding developments in the aerobic digestion process indicates that application of the process may increase as process kinetics, operational parameters, and design considerations are further defined.

*P*ROCESS THEORY

Aerobic digestion is based on the biological principle of endogenous respiration. Endogenous respiration occurs when the supply of available substrate (food) is depleted and the microorganisms begin to consume their own protoplasm to obtain energy for cell maintenance reactions.

During the digestion process, cell tissue is oxidized aerobically to carbon dioxide, water, and ammonia or nitrates. Because the aerobic oxidation process is exothermic, a net release of heat occurs during the process. Although the digestion process should theoretically go to completion, in actuality only 75 to 80% of the cell tissue is oxidized. The remaining 20 to 25% is composed of inert components and organic compounds that are not biodegradable. The material that remains after the full completion of the digestion process exists at such a low energy state that it is essentially biologically stable. Consequently, it is suitable for a variety of disposal options.

The aerobic digestion process actually entails two steps: the direct oxidation of biodegradable matter, and the subsequent oxidation of microbial cellular material by organisms. These processes are illustrated by the following formulas (U.S. EPA, 1979):

$$\text{Organic matter} + NH_4^+ + O_2 \xrightarrow{\text{Bacteria}} \text{Cellular material} + CO_2 + H_2O \tag{2.1}$$

$$\text{Cellular material} + O_2 \xrightarrow{\text{Bacteria}} \text{Digested sludge} + CO_2 + H_2O + NO_3^- \tag{2.2}$$

Equation 2.1 describes the oxidation of organic matter to cellular material. This cellular material is subsequently oxidized to digested biosolids. The process represented by Equation 2.2 is typical of the endogenous respiration process and is the predominant reaction in aerobic digestion systems.

Because of the need to maintain the process in the endogenous respiration phase, aerobic digestion typically is used to stabilize waste-activated sludge. The inclusion of primary solids in the process can shift the overall reaction to Equation 2.1, because they contain little cellular material. Most of the organic and particulate material in primary sludge comprises an external food source for the active biomass contained in the biological sludge. Therefore, longer retention times are required to accommodate the metabolism and cellular growth that must occur before endogenous respiration conditions are achieved.

Using the formula $C_5H_7NO_2$ as representative of the cell mass of a microorganism, the stoichiometry of the aerobic digestion process can be represented by either of the following equations:

$$C_5H_7NO_2 + 5O_2 \rightarrow 5CO_2 + 2H_2O + NH_3 + \text{Energy} \qquad (2.3)$$

$$C_5H_7NO_2 + 7O_2 \rightarrow 5CO_2 + 3H_2O + NO_3 + H^+ + \text{Energy} \qquad (2.4)$$

Equation 2.3 represents a system designed to inhibit nitrification; nitrogen appears in the form of ammonia. The stoichiometry of a system in which nitrification occurs is represented by Equation 2.4 where nitrogen appears in the form of nitrates.

Theoretically, approximately 50% of the alkalinity consumed by nitrification can be recovered by denitrification. If excessive pH depression is a problem (as a result of alkalinity consumption by nitrification), it may be possible to control this problem by periodic denitrification or by the addition of lime. Denitrification could be accomplished by periodically turning off the aerators while continuing to mix the digester if the facility is designed with a draft tube aerator containing an air sparger.

As indicated by Equation 2.4, nitrification during the aerobic digestion process increases the concentration of hydrogen ions and, subsequently, decreases pH if there is insufficient buffering capacity in the sludge. As in the activated-sludge process, approximately 7 kg of alkalinity are destroyed per kilogram of ammonia oxidized (7 lb/lb). The pH may drop as low as 5.5 during long aeration times; however, the aerobic digestion process does not seem to be adversely affected.

Equations 2.3 and 2.4 indicate that, theoretically, 1.5 kg of oxygen are required per kilogram of active cell mass (1.5 lb O_2/lb) in the nonnitrifying system, whereas 2 kg O_2/kg (2 lb O_2/lb) are required when nitrification occurs. Actual oxygen requirements for the aerobic digestion process depend on factors such as operating temperature, inclusion of primary solids, and the solids retention time in the activated-sludge system.

*P*ROCESS DESIGN

GENERAL. This discussion regarding the design of aerobic digestion systems is directed toward conventional aerobic systems—those systems operating at temperatures between 20 and 30°C that use air as the oxygen source for biological activity.

Factors that govern the design of aerobic digestion systems include desired reduction in volatile solids; quantities and characteristics of the influent; process operating temperature; oxygen transfer and mixing requirements; tank volume/detention time; and method of system operation.

REDUCTION IN VOLATILE SOLIDS. The primary purpose of the aerobic digestion process is to produce biosolids that are stabilized and amenable to various disposal options. Stabilized implies that the biological organisms, particularly pathogenic organisms, have been reduced to a level where no significant adverse environmental impact results from the disposal of the biosolids.

Reductions ranging from 35 to 50% in the levels of volatile solids are obtainable through the aerobic digestion process. The 40 CFR Part 503 regulations require that a 38% volatile solids reduction be met to attain the vector reduction requirements. The regulations governing land application of biosolids classify aerobic digestion as a process to significantly reduce pathogens that can produce biosolids suitable for land application.

Researchers have, however, suggested that the reduction in volatile solids during the digestion process may not be a valid indication of stabilization (Hartman *et al.*, 1979, and Matsch and Drnevich, 1977). Other parameters, such as the residual rate of oxygen demand, pathogen levels, odor-producing potential, specific oxygen uptake rate, or oxidation–reduction potential, may be more indicative of stabilized aerobically digested biosolids.

FEED QUANTITIES/CHARACTERISTICS. The aerobic digestion process typically is used to stabilize biological solids, such as waste-activated sludge. The process has been used to stabilize primary and biological mixtures; however, the associated retention times and oxygenation requirements are substantially increased to obtain a stabilization level equal to that achieved with biological solids. Because the mechanism of the aerobic digestion process is similar to the activated-sludge process, the same concerns regarding variations in influent characteristics and levels of biologically toxic materials apply, although a dampening effect will occur as a result of upstream treatment processes. The accumulation of heavy metals in activated sludge by precipitation/adsorption that can occur at a pH greater than 7 may result in toxicity in the aerobic digester from resolubilization of the heavy metals under low pH conditions present in the digestion process.

The influent concentration is important in the design and operation of an aerobic digestion process. While maximizing feed concentrations by prior thickening processes requires higher oxygen input levels per digester volume, thickening will result in longer solids retention times, smaller digester volume requirements, easier process control (less decanting), and subsequently increased levels of volatile solids destruction.

OPERATING TEMPERATURE

The operating temperature of the aerobic digestion system is the critical parameter in the process. A frequently cited disadvantage of the aerobic process is the variation in process efficiency that results from changes in operating temperature. Changes in operating temperatures are closely related to ambient temperatures because most aerobic digester systems use open tanks.

Aerobic digestion systems typically are operated in the mesophilic zone of bacterial action, while the operating temperature ranges from 10 to 40°C. There has been increased research into the operation of aerobic systems in the other temperature zones—the cryophilic zone of less than 10°C and the thermophilic zone of more than 40°C.

Because aerobic digestion is a biological process, the effects of temperature can be estimated by the following equation:

$$(K_d)_T = (K_d)_{20°C}\theta^{\,T-20} \tag{2.5}$$

Where

K_d	=	reaction rate constant, time
θ	=	temperature coefficient; and
T	=	temperature, °C.

The reaction rate constant indicates the destruction rate of volatile suspended solids during the digestion process. An increase in the reaction rate constant typically occurs with an increase in the temperature of the system and implies an increase in the digestion rate. Temperature coefficients ranging from 1.02 to 1.10 have been reported, with 1.05 as the average.

The rate of biological processes in the mesophilic temperature range typically increases with temperature (Figure 2.1). Above a critical temperature, the process will be inhibited. One study showed a maximum volatile solids destruction rate at 30°C, with a rate reduction at higher temperatures (Hartman *et al.*, 1979). This is in contrast with the data in Figure 2.1, and indicates the importance of obtaining rate data applicable to the system being designed.

OXYGEN TRANSFER AND MIXING REQUIREMENTS. The biological reaction that occurs during the aerobic digestion process requires oxygen for the respiration of cellular material in activated sludges and, in the case of mixtures with primary solids, the oxygen needed to convert organic matter to cellular material. In addition, proper operation of the system requires adequate mixing of the contents to ensure proper contact of oxygen, cellular material, and organic matter (food source). Because the introduction of oxygen to maintain the biological process typically provides a mixing action, these parameters are interrelated.

In aerobic digestion systems that treat strictly biological solids, the need for adequate mixing typically will govern the capacity of the oxygenation equipment. Systems treating primary and biological mixtures require additional oxygen for the biological oxidation process and, in many cases, this requirement will govern mixing equipment size.

Actual mixing requirements typically range from 10 to 100 W/m^3 (0.5 to 4.0 hp/1 000 cu ft) of digester volume; however, this value will vary depending on the type of mixing device. An experienced equipment manufacturer should be consulted to determine actual mixing requirements.

Equations 2.3 and 2.4 indicate that 1.4 to 1.98 parts of oxygen per part of applied organic cell mass are theoretically required for the aerobic digestion process, depending on whether nitrification is inhibited or allowed to proceed. Design experience has shown that 2.0 parts of oxygen per part of organic cell mass destroyed represents a standard minimum value for biological stabilization. The inclusion of primary solids in the digestion process requires an additional 1.6 to 1.9 parts oxygen per part of volatile solids destroyed to convert the organic matter to cell tissue and satisfy the endogenous demand of the resulting cell mass. Some U.S. state regulatory agencies, such as the state of Wisconsin, stipulate in their design standards for wastewater treatment plants (WWTPs) that the aeration system should account for an additional

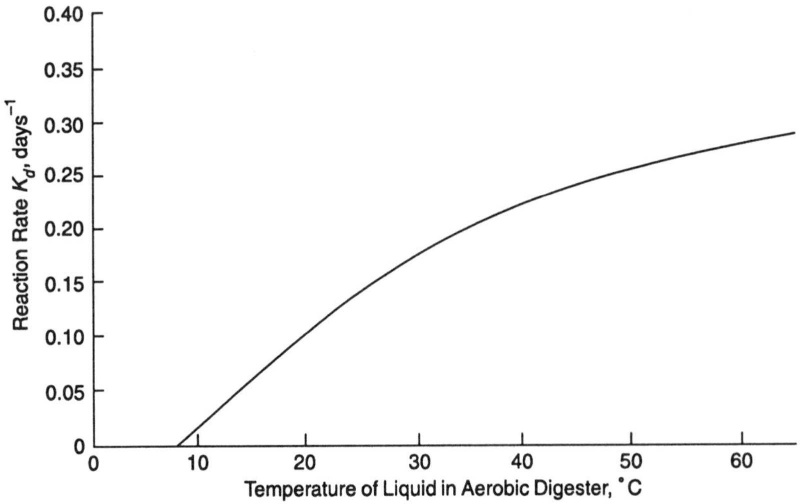

Figure 2.1   Reaction rate K_d versus aerobic digester liquid temperatures.

2 lb. of oxygen per pound of BOD_5 applied by primary sludge in the aerobic digestion design.

Aerobic digestion is a continuation of the suspended growth or fixed film biological treatment process; proper sizing of oxygen transfer equipment for the aerobic digestion system must reflect the reduction in oxygen demand that occurs in the biological process. In addition, oxygen requirements for aerobic digestion are increased if the secondary system is nitrifying. Estimating the oxygen requirement because of nitrification must be reflected not only in the ammonium concentration of the raw wastewater, but also in the conversion of organic nitrogen to ammonium in the secondary process and in primary sludge.

The oxygen requirements for an aerobic digester treating biological solids can be approximated from the following equation (WEF, 1992):

$$R_i = K(1.67\, S_i - O_a) \tag{2.6}$$

Where

R_i	=	actual aerobic digester oxygen requirements, kg/d (lb/d);
S_i	=	raw wastewater BOD_5 load equivalent to the waste solids under oxidation, kg/d (lb/d);
O_a	=	oxygen consumed in main aeration basin, kg/d (lb/d); and
K	=	constant equal to 1.0 when main aeration does not nitrify and 1.24 when main aeration does nitrify.

Oxygen requirements for aerobic digestion systems typically represent air flow rates of 0.25 to 0.33 $L/m^3 \cdot s$ (15 to 20 scfm/1 000 cu ft) for waste-activated sludges. Air flow rates increase to a range of 0.40 to 0.50 $L/m^3 \cdot s$ (25 to 30 scfm/1 000 cu ft) for a mixture of primary and waste-activated sludge (Benefield and Randall, 1980). Dissolved oxygen levels in the aerobic digester typically are maintained at approximately 2 mg/L; however, this level may be reduced if the oxygen uptake rate is less than 20 $mg/L \cdot h$.

After the requirements for adequate mixing and oxygen transfer have been computed separately, the larger of the two requirements will govern the overall system design. In cases where the mixing requirement exceeds the oxygen-transfer requirement, the design engineer should consider providing supplemental mechanical mixing rather than overdesigning the oxygen-transfer system. The increased capital cost for supplemental mechanical mixers must be balanced against power and maintenance costs of more aeration to determine the optimum configuration.

TANK VOLUME AND DETENTION TIME REQUIREMENTS. The required volume of an aerobic digestion system typically is governed by the detention time necessary to achieve a desired reduction in volatile solids. In the past, the detention time required to achieve a 40 to 45% reduction in volatile solids typically ranged from 10 to 12 days at an operating temperature of approximately 20°C (Metcalf and Eddy, Inc., 1979). The 40 CFR Part 503 regulations discuss a 38% volatile solids reduction that must be achieved in the stabilization processes; however, more importantly, the new designs must focus on pathogen reduction as the controlling factor. Although the destruction of volatile solids will continue with longer detention times, the oxidation rate decreases significantly and continuation of the digestion process past the typical detention time is not economical. It has been shown in full-scale aerobic digestion studies that a total aeration time (including time in the extended aeration process) of 35 to 50 days was required to consistently meet the 40 CFR Part 503 vector attraction reduction (503.33) for the specific oxygen uptake rate (SOUR) requirement of less than 1.5 mg O_2/g·h volatile solids at WWTPs located at higher altitudes with colder climates (Maxwell *et al.*, 1992). The total required aeration time was significantly dependent on operating temperature and biodegradability of the waste-activated sludge. It has been reported that a significant deterioration occurs in dewaterability of aerobically digested biosolids in systems with unusually long detention times (U.S. EPA, 1979). However, before the 40 CFR Part 503 regulations, many state agencies adopted U.S. EPA's standards for aerobic digestion that required a residence time of 60 days at 15°C and 40 days at 20°C and focused on pathogen reduction rather than on volatile solids reduction.

The reduction in biodegradable sludge during the digestion process typically is described by a first-order biochemical reaction at constant volume conditions, similar to the following:

$$\frac{dM}{dt} = K_d M \qquad (2.7)$$

Where

dM/dt	=	rate of change of biodegradable volatile solids per unit of time, Δ mass/volume/time;
K_d	=	reaction rate constant, time^{-1}; and
M	=	concentration of biodegradable volatile solids remaining at time, *t* (mass/volume).

The time factor in Equation 2.7 represents the solids retention time in the aerobic digester. Factors such as the method of digester operation, operating temperature, and the solids retention time of the activated-sludge system may result in the time factor being equal to, or greater than, the theoretical hydraulic residence time in the system. Use of the biodegradable portion of volatile solids in the equation recognizes that approximately 20 to 35% of the waste-activated sludge from WWTPs with primary treatment systems is nonbiodegradable. The percentage of nonbiodegradable volatile solids for waste-activated sludges from contact stabilization processes (no primary tanks) ranges from 25 to 35%.

The reaction rate constant, K_d, is a function of the type of residuals being digested, operating temperature, solids retention time of the system, and solids concentration in the digestion system. Figure 2.1 is a graph of the change in the reaction rate constant versus increasing operating temperature. The results of one study with waste-activated sludge at a temperature of 20°C indicated that the reaction rate constant declined with an increase in the suspended solids level in the digester (Reynolds, 1973).

The product of temperature and solids retention time appears to correlate with the percentage of volatile solids destruction that can be achieved during digestion. The selection of a desired percent reduction in volatile solids, coupled with an assumed operating temperature for the system, can be used to estimate the required digester solids retention time.

As the temperature and residence time in the digester increase, the fractional increase in volatile solids reduction declines. Based on one set of pilot plant results, one research team has concluded that comparatively little additional reduction in volatile solids occurs when this product exceeds 250 in the operating temperature range of 10 to 20°C. In the U.S. the final regulations (40 CFR Parts 257 and 503) require that a 38% volatile solids reduction occur in the aerobic digestion process based on vector attraction reduction objectives. This reduction in the aerobic digester may be difficult to obtain if significant biodegradable solids reduction has already occurred in a secondary treatment system with a long solids retention time. In the 1980s, design engineers were sometimes allowed variances from the state regulatory agencies in which credit was given for the volatile solids reduction that occurred in the extended aeration facilities. Therefore, the overall volatile solids reduction took into account the waste flow through the activated-sludge process and the solids-handling process. This variance allowed municipalities to meet 38% volatile solids reduction through their facilities. With the enactment of 40 CFR Part 503, state regulatory agencies no longer allow design engineers to count the reduction of volatile solids through the secondary treatment process as part of the 38% that is to occur in the aerobic digestion process or other stabilization processes. If the 38% reduction is not attained in the aerobic digester, the regulations have stated that vector attraction reduction can be demonstrated by digesting a portion of the previously digested material that has a solids concentration of 2% or less aerobically in the laboratory in a bench-scale unit for 30 additional days at 20°C. When at the end of the 30 days, the volatile solids from the beginning of that period are reduced by less than 15%, vector attraction reduction is achieved. In addition, based on the 40 CFR Part 503 regulations, design engineers and state regulatory agencies are using the SOUR criteria of < 1.5 mg O_2/g·h total solids at a temperature of 20°C for digested biosolids instead of the 38% volatile solids reduction.

One method to determine the volume of a continuously operating aerobic digester is to apply the following formula:

$$V = \frac{Q_i(X_i + YS_i)}{X(K_dP_v + 1/SRT)} \tag{2.8}$$

Where

V	=	volume of the aerobic digester, L (cu ft);
Q_i	=	influent average flow rate, L/d (cu ft/d);
X_i	=	influent suspended solids, mg/L;
Y	=	portion of the influent BOD consisting of raw primary solids, %;
S_i	=	influent digester BOD_5, mg/L;
X	=	digester suspended solids, mg/L;
K_d	=	reaction rate constant, d^{-1};
P_v	=	volatile fraction of digester suspended solids, %; and
SRT	=	solids retention time, d.

The term YS_i in Equation 2.8 can be disregarded if no primary solids are included in the load to the aerobic digester. Equation 2.8 should not be used to compute digester volumes in systems where significant nitrification will occur.

Research by Benefield and Randall (1980) has resulted in the development of equations for determining required digester detention times. These proposed equations result from an analysis of the kinetics associated with the digestion process plus the understanding that a portion of the volatile solids in the process are nonbiodegradable and a portion of the nonvolatile solids are solubilized from the lysis of the microbial cells contained within the sludge. The basic equation is as follows:

$$t_d = \frac{(X_i - X_e) + (Y)(S_a)}{(K_d)(D)(X_{oad})(X_i)} \tag{2.9}$$

Where

t_d	=	digester detention time, d;
X_i	=	TSS concentration in the influent, mg/L;
X_e	=	TSS concentration in the effluent, mg/L;
X_{oad}	=	percentage of active biomass that is biodegradable in the influent;
K_d	=	reaction rate constant for the biodegradable portion of the active biomass, d^{-1};
D	=	biodegradable active biomass in the influent that appears in the effluent, %;
Y	=	yield coefficient for organic content of the primary solids; and
S_a	=	ultimate BOD of the primary solids in the digester effluent, mg/L.

Wastewater Residuals Stabilization

If the aerobic digestion system is treating only biological solids, the factors Y and S_a are equal to zero and the other parameters are representative of the waste-activated sludge. Treatment of a mixture of primary and waste-activated sludge requires the inclusion of the Y and S_a factors to reflect the conversion of organic matter to cellular material. The inclusion of additional factors for the primary component is based on the premise that the presence of these solids slows the destruction rate of cellular material, while the organic matter is converted to cellular material. The other terms in the equation should reflect the characteristics of the mixture.

Equation 2.9 supports the assumption that for equivalent solids reduction and constant solids loading, solids retention time must be increased as the active fraction of the influent biomass decreases. Actual operating experience indicates that for systems with a very low fraction of active biomass in the feed, which is typical for extended aeration systems, this trend does not hold. For these systems, the detention times computed by Equation 2.9 may be reduced.

SUMMARY OF DESIGN PARAMETERS. The enactment of the 40 CFR Part 503 regulations has substantially changed the typical design parameters used for the aerobic digestion process. For example, on early aerobic digester designs that were receiving only waste-activated sludge, the design engineers would use a residence time of 10 to 15 days. Then U.S. EPA focused on pathogen reduction and set forth regulations in which the residence time had to be 40 days at a temperature of 20°C and 60 days if the wastewater temperature was 15°C. The 40 CFR Part 503 regulations require residence times only if the designer wants to meet the Class B pathogen reduction criteria without the need to periodically monitor for fecal coliform. If the design engineer uses the aerobic digestion process for stabilization and does not meet the 40 days at 20°C or the 60 days at 15°C, then the design engineer must monitor. In both cases the design engineer must demonstrate that the vector attraction has been achieved by a 38% volatile solids reduction, an oxygen uptake rate of < 1.5 mg O_2/g·h total solids or a volatile solids reduction of less than 15% on further testing after 30 days of additional digestion has been achieved. If the design engineer needs the 40 days at 20°C or 60 days at 15°C (or a linear interpolation between them) then the design engineer does not have to monitor fecal coliform, but must show vector attraction reduction.

DESIGN CONSIDERATIONS

METHOD OF OPERATION. The two primary operational modes for aerobic digestion systems are batch operation or continuous operation. The major difference between the batch and continuous modes is the manner in which supernatant is withdrawn from the process. Figure 2.2 illustrates operational modes.

In the batch mode of operation, the aerobic digestion tanks are fed directly from the clarifiers. The filling operation continues for a period of time depending on wasting rates, precipitation, and evaporation. The oxygenation/mixing equipment operates during the filling period that typically occurs for 1 to 2 days, or in extreme cases for as long as 2 to 3 weeks. After the tank is full the oxygenation/mixing must continue until the stabilization process is completed by following the pathogen reduction and vector attraction requirements stipulated in 40 CFR Part 503.

After the stabilization process cycle has completed the oxygenation/mixing equipment is turned off and the biosolids are allowed to settle. A supernatant layer forms at the surface and is subsequently removed to create additional volume for the introduction of more feed. Digested biosolids are removed from the tank on a regular basis; however, some biosolids are always left in the tank to serve as a seed for the next batch. The frequency of supernatant and digested biosolids removal is determined by the feed requirement

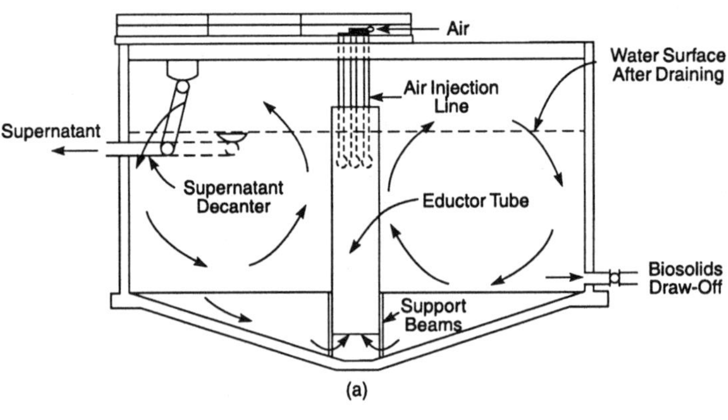

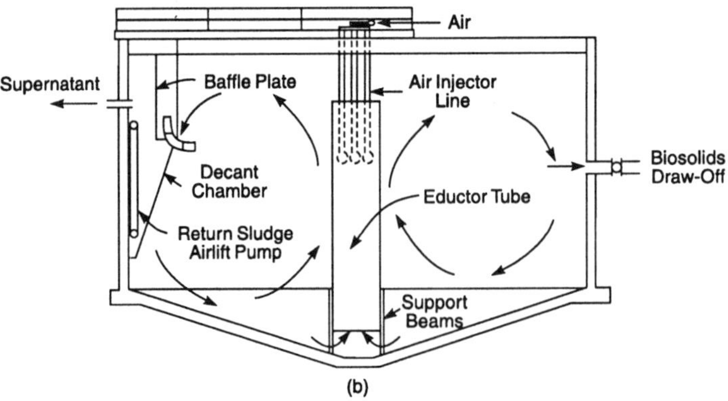

Figure 2.2   Typical circular aerobic digester configurations: (a) batch operation and (b) continuous operation (WEF, 1992).

and the desired solids retention time. Multiple supernatant draw-off points should be provided for flexibility in operation. Batch operation typically is selected for small-capacity digestion systems because of the relative simplicity of operation.

The continuous mode allows regular operation without interrupting the oxygenation/mixing equipment. The decanting, filling, and aeration operation is a continuous process. Baffles and a supernatant withdrawal chamber (stilling well) or separate settling tank are common features of the continuous mode of operation. Including stilling wells in the digestion tank typically is unacceptable because of the mixing-induced turbulence carried into the stilling well (WEF, 1992). Floating solids may result from denitrification occurring in stilling wells and contribute to occasional poor supernatant quality.

The design of separate settling basins for continuous mode aerobic digestion is similar to the design of gravity thickeners or flotation thickeners for waste-activated sludge. Solids loading rates ranging from 25 to 50 kg/m^2·d (5 to 10 lb/d/sq ft) of gravity thickener surface typically are used. Higher rates of 50 to 100 kg/m^2·d (10 to 20 lb/d/sq ft) are allowable with flotation thickeners. Provisions should be included for rapid biosolids withdrawal and baffling/skimming to remove any accumulated floating solids. Thickened biosolids withdrawn from a batch digestion process or the continuous digestion process typically ranges from 2 to 4% solids with gravity thickeners and 4 to 6% solids from flotation thickeners or gravity belt thickeners.

TANK DESIGN. Aerobic digesters typically are uncovered, unheated aeration tanks of steel or concrete construction; however, several recent installations have included covered tanks to reduce the temperature-dependent nature of the process. Design factors used to determine the required volume for aerobic digestion systems should be carefully evaluated to avoid creating increased energy requirements for mixing and increased heat loss from the system that can result from overdesigned tank volumes.

At least two tanks should be provided in most aerobic digestion systems to permit draining and equipment repair. Multiple units are especially important in batch operation systems to provide digester capacity during the supernatant formation cycle. Aerobic digestion tanks have been designed with rectangular, circular, or annular geometry. Bottom slopes in the tanks typically range from 1:12 to 3:12 to facilitate biosolids removal. Side water depths are similar to those provided for activated-sludge systems, but more generous freeboards typically are provided (1 to 3 m [3 to 10 ft]) to contain excessive foaming that may occur.

AERATION EQUIPMENT. Several types of aeration devices have been used successfully to provide the oxygenation and mixing requirements of aerobic digestion systems. These devices are classified as diffused air, mechanical surface aeration, or mechanical submerged turbines.

The design of diffused air systems for aerobic digesters is comparable to those used in standard activated-sludge systems. Diffusers typically are located near the tank bottom and along one tank side to produce a spiral or cross-roll pattern, although floor-mounted grid systems can also be used. Air flow rates of 0.25 to 0.33 $L/m^3 \cdot s$ (15 to 20 cfm/1 000 cu ft) typically are needed for oxygenation requirements; however, 0.33 to 0.67 $L/m^3 \cdot s$ (20 to 40 cfm/1 000 cu ft) typically are required to ensure adequate mixing. While both fine-bubble and coarse-bubble diffusers have been used in aerobic digesters, the nonclog- and porous-media-type devices are more resistant to plugging than the large-bubble, orifice-type diffuser (WEF, 1992). However, surface fouling of porous diffusers may occur.

Diffused air systems have the following advantages: oxygen transfer can be controlled by varying the air supply rate; the introduction of compressed air into the digester typically adds heat to the system with a resultant increase in the rate of biological activity; diffused air systems are less prone to the creation of severe foaming problems that can result with other devices; and overall heat loss from the system is reduced because of the relatively small degree of surface turbulence.

These advantages of diffused air systems are frequently outweighed by the recurring clogging problems that occur in aerobic digesters (particularly in batch systems) and the variations in oxygen transfer and mixing efficiencies that occur as a result of fluctuations in tank levels during fill and draw operations. If a diffused air system is to be used, it is imperative that provisions be included for the removal of a digester so the diffusers can be cleaned or for easy removal of the diffuser device and air drop pipes for regular cleaning.

Although a diffused air system adds heat to the digester from the compressed air system, some heat loss can still occur because of evaporation caused by the large volumes of gas that must be used. The air passing through the digester can strip water vapor and heat (equivalent to the latent heat of vaporization) from the sludge. It has been calculated that autothermal digestion cannot take place if air is used as the aeration gas, because of the excessive heat loss (Gould and Drnevich, 1978).

Mechanical surface aerators have relatively high oxygen-transfer efficiencies and typically are floating, pontoon-mounted devices of either low- or high-speed design. The low-speed design is used more often in aerobic digestion systems. Mechanical surface aerators provide an efficient, low-maintenance method for supplying air to the digestion process. The disadvantages attributed to surface aeration include the lack of control of the oxygenation rate; performance deterioration if excessive foam is present; an increased potential for foaming problems because of high surface turbulence; and increased heat loss from the system and the potential for ice accumulations during the winter, as a result of the splashing action of the device. Floating aerators also are difficult to maintain, particularly if no permanent access to the aerators is provided.

Mechanical submerged turbine aerators combine several advantages and eliminate some disadvantages of the diffused air and surface aerator devices. Oxygenation rates can be controlled by varying the air flow rate to the submerged impeller. Because the impeller is submerged, these devices are not as sensitive to foaming conditions and do not have the ice and heat dissipation problems associated with surface aerators.

*O*PERATIONAL CONSIDERATIONS

One of the advantages of the aerobic digestion process compared to anaerobic systems is its relatively simple operation. As long as a standard biological environment (temperature and pH control, exclusion of toxic substances, and so forth) is maintained comparable with an activated-sludge system, the aerobic digestion process is essentially self-sustaining.

SUPERNATANT QUALITY. Both the batch and continuous modes of operation result in the supernatant being discharged back to the WWTP. While the quality and quantity of this supernatant and its effect on the WWTP must be assessed, the characteristics of the supernatant are relatively innocuous. This constitutes one of the advantages of the aerobic as compared with the anaerobic digestion system.

Table 2.1 presents ranges and typical values for various constituents of the supernatant discharged from aerobic digestion systems. It should be noted that the suspended solids in the supernatant may be high in concentration; however, the largest fraction of these suspended solids comprises stabilized biosolids, and the recycle of the supernatant to the headworks of the WWTP typically will result in biosolids removal in the primary settling process (U.S. EPA, 1974).

Table 2.1   Characteristics of supernatant from aerobic digestion systems.

Parameter	Range	Typical value
pH	5.9–7.7	7.0
BOD_5, mg/L	9–1 700	500
Filtered BOD_5, mg/L	4–183	50
COD, mg/L	288–8 140	2 600
Suspended solids, mg/L	46–11 500	3 400
Kjeldahl nitrogen, mg/L	10–400	170
NO_3–N, mg/L	—	30
Total phosphorus, mg/L	19–241	100
Soluble phosphorus, mg/L	2.5–64	25

FOAMING PROBLEMS. Foaming problems in aerobic digestion systems are not uncommon, particularly in systems using surface aeration devices. These foaming problems typically are caused by high organic loading rates during warm weather periods. Water sprays typically are used to control the problem. Antifoaming chemicals may be added to the spray water in response to more severe foaming problems.

Growths of filamentous bacteria can also cause foaming problems. Control methods that have yielded mixed results include adding oxidizing chemicals to destroy the filamentous bacteria, and creating temporary anaerobic conditions by turning off the aeration equipment in the digester (WEF, 1990).

Foaming may also occur in the spring and fall as the microorganisms reacclimate to summer or winter temperatures.

pH REDUCTION. If the aerobic digestion process is provided with sufficient oxygen and detention time, ammonia will nitrify and form nitrates. This nitrification process may result in a decrease in pH and alkalinity as a result of the acid formed during the process. If the aerobic digester has separate mixing and aeration equipment, it is possible to denitrify by operating only the mixing equipment during the fill cycle. Although the pH reduction should be monitored and the buffering capacity of the system checked as a normal operational control parameter, it seems that the organisms in the aerobic digestion process will acclimate to the lower pH as long as the level does not drop suddenly and does not continue to decrease below a pH of approximately 5.5 (Metcalf and Eddy, Inc., 1979).

DEWATERING. The literature addressing the dewatering characteristics of aerobically digested biosolids is extensive, but occasionally contradictory. Although one of the frequently cited advantages of the aerobic digestion process is the dewatering characteristics of the stabilized biosolids, recent research has indicated that the mechanical dewatering of aerobically digested biosolids is difficult. More importantly, the dewatering characteristics show a definite deterioration with increasing solids retention time. Optimal dewaterability seems to occur after 1 to 5 days of aeration with a marked deterioration in dewaterability with increased aeration times (U.S. EPA, 1979).

The dewaterability of aerobically digested biosolids is also affected by the degree of mixing provided during the digestion process—high degrees of mixing destroy the structure of the solids floc with a subsequent adverse effect on dewaterability. Consequently, the design of mechanical dewatering facilities following aerobic digestion should incorporate conservative criteria developed during field testing and consultation with representatives of experienced equipment manufacturers.

*P*ROCESS VARIATIONS

GENERAL. Several variations to the standard mesophilic-air oxygenation aerobic digestion system have been investigated in recent years. The more notable variations are high-purity oxygen aeration, autothermal thermophilic digestion, and low-temperature digestion.

HIGH-PURITY OXYGEN AERATION. This modification of the aerobic digestion process substitutes high-purity oxygen for air. High-purity oxygen systems are relatively insensitive to changes in ambient air temperatures because of the increased rate of biological activity and the exothermic nature of the process.

While one variation of this modification uses open tanks, aerobic digestion using high-purity oxygen typically is done in closed tanks similar to those used in the high-purity oxygen activated-sludge process. The use of the high-purity oxygen aerobic digestion system in enclosed tanks typically will result in higher operating temperatures because of the exothermic nature of the digestion process.

The major disadvantage of this modification is the increased cost associated with oxygen generation. As a result, high-purity oxygen aerobic digestion typically is cost effective only when used in conjunction with the comparable activated-sludge process. The use of high-purity oxygen also decreases the amount of carbon dioxide that would be added in the standard air-oxygenated system. Consequently, neutralization may be required to offset the reduced buffering capacity of the system.

THERMOPHILIC AEROBIC DIGESTION. Extensive research has been conducted optimizing the exothermic nature of the aerobic digestion process by containing the released heat and operating the process in the thermophilic zone of biological activity. This modification is known as autothermal thermophilic aerobic digestion. The advantages of thermophilic operation are the decrease in retention times required to achieve a given suspended solids reduction and the production of an essentially pathogen-free end product as a result of the high degree of pathogen kill that occurs at the higher operating temperature.

Operation in the thermophilic range results in nitrification inhibition. Therefore, less oxygen is required than for systems in which nitrification occurs.

As a result of the reduction in the aeration gas volume required, the normal heat loss from the aerobic digestion system is reduced. This reduction results in a retention of heat quantities generated during the biological stabilization process. Covering and insulating the digestion tanks increases the quantity of heat retained, and helps to maintain the process in the thermophilic range.

The autothermal thermophilic aerobic digestion process is self-regulating with respect to temperature. This occurs as a result of a decrease in the digestion rate at elevated temperatures. The decrease in biological activity reduces the quantity of heat released during the exothermic reaction with a resultant decrease in the process operating temperature. The process is relatively stable, recovers quickly from minor process upsets, and is not affected by relatively wide variations in outside air temperature.

LOW-TEMPERATURE AEROBIC DIGESTION. The operation of aerobic digestion systems at lower temperature ranges (less than 20°C) has been studied to provide better operational control for small package-type WWTPs in northern climates. Investigations by researchers at WWTPs in British Columbia have indicated that the solids retention time must be increased as operating temperatures decrease to maintain an acceptable level of suspended solids reduction (Koers and Mavinic, 1977, and Mavinic and Koers, 1979). Research determined that the product of solids retention time (days) and operating temperature (°C) should be maintained in the range of 250 to 300 days for operating temperatures between 5 and 20°C to ensure acceptable volatile solids reductions.

R*EFERENCES*

Benefield, L.D., and Randall, C.W. (1980) *Biological Process Design for Wastewater Treatment.* Prentice-Hall, Englewood Cliffs, N.J.

Gould, M.S., and Drnevich, R.F. (1978) Automated Thermophilic Aerobic Digestion. *J. Environ. Eng.,* **104,** 259.

Hartman, R.B., *et al.* (1979) Sludge stabilization through aerobic digestion. *J. Water Pollut. Control Fed.,* **51,** 2353.

Koers, D.A., and Mavinic, D.S. (1977) Aerobic digestion of waste activated sludge at low temperatures. *J. Water Pollut. Control Fed.,* **49,** 460.

Matsch, L.C., and Drnevich, R.F. (1977) Autothermal aerobic digestion. *J. Water Pollut. Control Fed.,* **49,** 296.

Mavinic, D.S., and Koers, D.A. (1979) Performance and kinetics of low temperature, aerobic sludge digestion. *J. Water Pollut. Control Fed.,* **51,** 2088.

Maxwell, M.J., *et al.* (1992) Impact of New Sludge Regulations on Aerobic Digester Sizing and Cost-Effectiveness. Paper presented at 65th Annu. Conf. Water Environ. Fed., New Orleans, La.

Metcalf and Eddy, Inc. (1979) *Wastewater Engineering: Treatment, Disposal, Reuse.* McGraw–Hill, Inc., New York, N.Y.

Reynolds, T.D. (1973) Anaerobic Digestion of Thickened Waste Activated Sludge. *Proc. 28th Ind. Waste Conf., Purdue Univ.,* West Lafayette, Ind.

U.S. Environmental Protection Agency (1974) *Process Design Manual for Upgrading Existing Wastewater Treatment Plants.* EPA-625/1/71-004a, Technol. Transfer, Washington, D.C.

U.S. Environmental Protection Agency (1979) *Process Design Manual for Sludge Treatment and Disposal.* EPA-625/1-79-011, Technol. Transfer, Washington, D.C.

Water Environment Federation (1990) *Operation of Municipal Wastewater Treatment Plants.* Manual of Practice No. 11, Alexandria, Va.

Water Environment Federation (1992) *Design of Municipal Wastewater Treatment Plants.* Manual of Practice No. 8, Alexandria, Va.; Am. Soc. Civ. Eng., Manual and Report on Engineering Practice No. 76, New York, N.Y.

Chapter 3
Autothermal Thermophilic Aerobic Digestion

The autothermal thermophilic aerobic digestion (ATAD) process is an aerobic digestion process that achieves thermophilic operating temperatures (40 to 80°C) without supplemental heat beyond that supplied by mixing energy. Within the ATAD reactor, sufficient levels of oxygen, volatile solids, and mixing allow aerobic microorganisms to degrade organics to carbon dioxide, water, and nitrogen byproducts, during which heat energy is released. If sufficient insulation, hydraulic retention time (HRT), and adequate solids concentration are provided, the process can be controlled at thermophilic temperatures to achieve greater than 38% volatile solids destruction and pathogen reduction sufficient to meet U.S. Environmental Protection Agency (U.S. EPA) regulations for processes for further reduction of pathogens (PFRP) or the 40 CFR Part 503 Class A designation.

The ATAD process has been studied since the 1960s. Much of the developmental work was done by Popel (1971a and b) and co-workers on animal manure and wastewater residuals in the Federal Republic of Germany (FRG). Development of an aspirating aeration device was key to the FRG process success. Research in the U.S. was done by Matsch and Drnevich (1977) using pure oxygen, and by Jewell and Kabrick (1980) using air with submersible aeration devices.

*P*ROCESS THEORY

In an ideal aerobic digestion process, organic decomposition would result in the formation of carbon dioxide and water. Assuming all organics in the digester are cell mass and that the formula $C_5H_7NO_2$ is representative of the cell mass of a microorganism, the chemical changes brought about in an aerobic digester can be described by one of the following equations (Matsch and Drnevich, 1977):

$$C_5H_7NO_2 + 5O_2 \rightarrow 5CO_2 + 2H_2O + NH_3 + Energy \qquad (3.1)$$

or

$$C_5H_7NO_2 + 7O_2 \rightarrow 5CO_2 + 3H_2O + NO_3 + H^+ + Energy \qquad (3.2)$$

Equation 3.1 is indicative of a system operating at conditions designed to inhibit nitrification. In ATAD systems, operating temperatures of more than 40°C inhibit nitrification; therefore, Equation 3.1 describes the ideal reaction occurring in an ATAD system. This expression indicates that approximately 1.5 kg of oxygen is required per kg of volatile solids destroyed. The energy produced approximates 21 000 kJ/kg of volatile solids destroyed (Andrews and Kambhu, 1970).

Autothermal thermophilic aerobic digestion systems are designed to have a short HRT within insulated reactors. As long as the system is well mixed and sufficient oxygen is provided, the temperature in the reactor will rise until a balance occurs (heat lost equals heat input from exothermic reaction and mechanical energy input). The temperature continues to increase as the solids load to the system increases until the system becomes oxygen mass transfer limited. Typical operating temperatures in the final reactor are 60 to 65°C. Temperatures more than this amount are feasible and are limited by volatile solids content, mixing energy, oxygen supply, and heat loss.

The high temperatures achieved in the ATAD system offer several benefits:

- Reduced HRT (5 to 6 days) to achieve similar volatile solids reductions as conventional aerobic digestion (30 to 50% volatile suspended solids [VSS] reduction).
- Use of excess heat generated by the process for building heating or preheating the autothermal thermophilic aerobic digestion influent feed with a series of heat exchangers.
- Pasteurization, resulting in die off of pathogenic organisms.

*P*ROCESS DESIGN

To achieve an effective ATAD design, an integrated solids handling approach must be used to provide proper feed characteristics, a proper environment for the autothermal reaction, and postprocess biosolids cooling. These features can be integrated into existing facilities as a retrofit project or into new facilities designed specifically for the ATAD process. Key elements of this process are described below. A typical ATAD configuration is shown in Figure 3.1.

PRETHICKENING SYSTEM. To effectively operate with minimum mixing energy supplement, the feed must have a chemical oxygen demand (COD) of 40 g/L or greater. This is best achieved by thickening to achieve a minimum solids concentration of 3% before ATAD feed. The prethickener can also be used as a blend tank for blending primary and secondary solids. Prethickening can be accomplished by gravity or by mechanical means.

REACTORS. Typically, a minimum of two enclosed, insulated reactors are provided in series, and are provided with mixing, aeration, and foam control equipment. Single-stage systems can be provided and can achieve similar volatile solids reduction as multiple-stage systems, but unless the system is batch operated, pathogen reduction may be less because of short circuiting

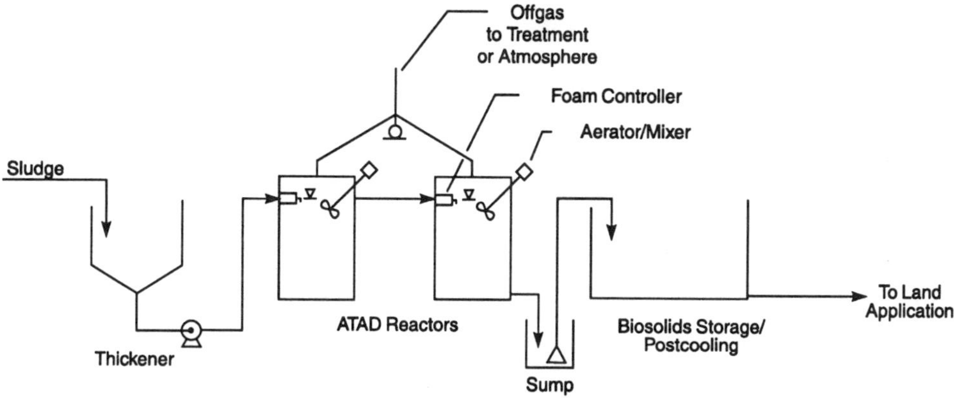

Figure 3.1   Typical autothermal thermophilic aerobic digestion system.

potential. Three- and four-stage systems can also be provided for additional process flexibility. Offgases are exhausted to the atmosphere or to treatment systems for odor reduction.

POSTCOOLING/THICKENING. After the ATAD process, biosolids cooling is sometimes necessary for effective thickening and enhanced supernatant quality. Twenty days of cooling is typically recommended. Other cooling systems using heat exchangers and reduced post ATAD detention times (1 to 3 days) are feasible.

FEED CHARACTERISTICS. Feed characteristics to the ATAD are critical to achieve successful operation. Typically, both primary and secondary solids are fed to the ATAD, although some facilities feed only waste secondary solids. Blending before feeding to the ATAD is optional.

A minimum feed solids concentration of 3% with a range of 4 to 6% and 40 g/L COD or greater is recommended. Solids concentrations less than 3% contain too much water and may not achieve autothermal conditions. Concentrations greater than 6% become difficult to effectively mix and aerate. Prethickening in Europe typically occurs by gravity in the primary clarifier or in a separate gravity thickener. Other thickening devices, such as rotary screens, dissolved air flotation devices, or gravity belt thickeners, may also be used.

The feed must contain a minimum volatile solids content of 25 g/L (2.5%) and 40 g/L COD. Volatile solids contents less than this amount will have difficulty achieving autothermal conditions. Feed from wastewater treatment plants (WWTPs) without primary clarifiers and operating with food-to-microorganism ratios (F:M) as low as 0.1:0.15 still seems suitable for the ATAD process. However, solids retention times of 15 days or fewer should be maintained so that volatile solids reduction via endogenous respiration in the aeration basins is kept to a minimum. The ATAD process can be used on low F:M feed solids but may require higher feed concentration or if the heat content of the feed is low, an external preheat source via heat exchangers within the reactors.

Either the raw wastewater or the solids should undergo fine screening (6- to 12-mm spacing [0.25- to 0.5-in. spacing]) to eliminate inerts, plastics, and rags from the reactor. Macerating has been used, but is not recommended as a replacement for screening. Where heat exchangers are used, maceration may be considered as a supplement to a good prescreening facility. Good grit removal from the wastewater is also necessary to reduce wear on aeration devices and grit buildup in the reactors.

DETENTION TIME. Initial testing in 1972 (Gay *et al.*, 1974) determined that a comparable degree of stabilization, as compared with aerobic digestion, could be achieved with the ATAD process at temperatures more than 45°C and HRT of 5 days or fewer. German design standards to comply with pathogen destruction requirements results in a 5- to 6-day HRT. Detention times of 4 to 30 days have been reported in the literature (Kelly, 1991a).

FEED CYCLE. Feed to the ATADs can occur either continuously or on a batch basis.

In the continuous mode, feed enters the first reactor continuously or semi-continuously. The contents of the first reactor are allowed to overflow to the second reactor and from the second reactor to a holding tank. In this mode, a means of maintaining constant levels in the reactors must be provided to achieve constant aerator submergence if aspirating aerators are used. Other aeration systems do not require strict level controls.

Batch feeding is sometimes designed to feed one day's volume to the reactor in less than 1 hour. Batch feeding results in 23 hours of retention time each day for pathogen destruction purposes. The batch feeding sequence may be described as follows for a two-stage system:

1. Shut off aerators and foam cutters in reactors #1 and #2.
2. Discharge biosolids from reactor #2 into holding tank.
3. Open valve between reactors and allow levels to equalize between reactors.
4. Add feed to reactor #1 until operating level is achieved, and allow contents to displace from reactor #1 to #2 while filling.
5. Close valve between reactors and turn on all aerators and foam cutters after reactors reach operating level.

AERATION AND MIXING. The key to effective ATAD performance is aeration and mixing. A highly efficient aeration/mixing device is required to

- Meet the process oxygen demand,
- Reduce heat loss that occurs from air exhausted from the reactors, and
- Provide adequate mixing to ensure complete stabilization.

Several types of aeration/mixing devices have been used with ATAD systems, including

- Aspirating aerators.
- Combination recirculation pump/Venturi arrangement, and
- Turbine and diffused air.

The most common device used is the aspirating aerator. Typical installations use a minimum of two aerators, side-mounted on each reactor. Larger installations may use a third center-mounted unit or additional tangentially mounted units. The angle of the aerator causes downward vertical mixing and

---

a circular horizontal flow pattern within the reactor. The advantage of this aerator design is that the motor and bearings are located outside the reactor.

Typical design parameters for this type of aeration device consist of the following:

- Specific power of 85 to 105 W/m^3 of active reactor volume,
- Air input of 4 $m^3/m^3{\cdot}h$ of active reactor volume,
- Energy requirement of 32 to 54 MJ/m^3 of solids throughput, and
- Oxygen transfer efficiency of 2 kg O_2/kWh.

The air input of 4 $m^3/m^3{\cdot}h$ of active reactor volume is premised on a feed VSS range of 2.5 to 5%. This value relates to a process oxygen requirement of 1.5 kg O_2/kg VSS destroyed at the aerator efficiency noted above.

A combination recirculation pump and Venturi arrangement has been used in the United Kingdom and Canada (Wolinski, 1985, and Kelly, 1991b). Research was conducted by the Water Research Center and Wessex Water Authority at the Palmersford Sewage Treatment Works, located in the United Kingdom.

Wolinski (1985) reports that compressed air was fed to the Venturi by a pressure regulator and gas flow meters. Both air and pure oxygen were tested. Air was found to be more effective than oxygen with nearly 100% use of the air introduced as opposed to 50 to 90% use of oxygen introduced. Correlation of this phenomenon with the foam layer showed improved air use with a thick foam blanket. When pure oxygen was used, the foam layer dissipated and oxygen use decreased. Therefore, the foam layer and oxygen use were interrelated. The main advantage of this type of aerator is that both the pump and Venturi are located outside of the digester. However, pump impeller wear was reported to be high and the potential exists for pump plugging. This points out the need for specifying the correct metallurgy for the pump internals.

A similar pump Venturi arrangement was used at the Haltwhistle ATAD plant in the United Kingdom (Murray *et al.*, 1990). Operational difficulties included pump wear from grit (poor grit removal) and pump clogging, even though a macerator was used.

A pump Venturi arrangement was used at Gibsons and at Whistler, British Columbia, Canada (Kelly, 1991a). At Gibsons, a 7-kW, variable-speed recirculation pump and Venturi are provided for each reactor, resulting in a power density of 675 W/m^3. Energy use is 290 MJ/m^3 of solids processed. This higher power density, when compared with an aspirating aerator design, was able to raise the reactor temperatures from 15 to 65°C in fewer than 3 days after a forced washout. The Whistler ATAD operates at 100 to 120 W/m^3 but is capable through a variable-frequency drive unit on the recirculation pump of providing 250 W/m^3 of power input during cold feed or low start-up temperatures. Pump plugging occurred at Gibsons until the pumps were replaced with solids-handling pumps. Since that time, no plugging has occurred. Use of corrosion-resistant impellers and pump volute lining is required. Erosion/corrosion cycling will wear mild steel quickly at high operating temperatures.

The turborator aerator is a type of aspirating aerator developed in Canada and used at the Salmon Arm, British Columbia, facility (Kelly, 1991a). A power density of approximately 190 W/m^3 was initially provided. When feed concentrations were less than 3 to 4%, washout conditions occurred, as evidenced by temperature drops. Subsequently, a third reactor was constructed to increase HRT to more than 10 days. The power density used for the system is now 105 W/m^3, with an energy consumption of 100 MJ/m^3 of solids processed.

TEMPERATURE AND pH. Process operating temperatures for a typical two-stage ATAD system are 35 to 50°C for reactor #1, and 50 to 60°C for reactor #2. An average temperature of 55°C is used in the second reactor for design purposes. The temperature in the second reactor typically exceeds 55°C and can go as high as 65 to 80°C without serious process problems.

During feeding, a drop in temperature occurs in the first reactor (the amount of drop depends on the quantity of feed and temperature) with a typical recovery rate of 1°C/h, if an aspirating aeration device is used. Actual temperature recovery rate depends on feed characteristics, inlet temperature, power density, and aerator efficiency. The temperature in the first stage should not be allowed to drop below 25°C to avoid biological adaptation problems. Canadian research has found that stabilization will be achieved at 40% reduction of COD or volatile solids with a two reactor degree day product of 400°C days or greater.

Typically, pH does not need to be controlled to affect process performance. German experience shows that with a feed pH of 6.5, values in the first reactor typically are near 7.2 and approach 8.0 in the second reactor.

FOAM CONTROL. Foam control plays an important role in the ATAD process. The foam layer is reported by Wolinski (1985) to affect oxygen-transfer efficiency and enhance biological activity. If left uncontrolled, an excessively thick, brown, foam layer forms and results in foam loss from the reactor. Foam control and development are necessary for effective operation.

Facilities may be designed with 0.5 to 1.0 m of freeboard, to use as volume for foam development and control. Control consists of breaking up large bubbles into smaller bubbles and densifying the layer. This can be accomplished by mechanical horizontal shaft foam cutters suspended in the reactor at fixed elevations. Air cooling is provided around the motor to enhance motor life. Other foam control methods include vertical mixers and spray systems. Selection of a system depends on reactor geometry and mixing patterns. Chemical defoamers are also used.

POSTTHICKENING AND DEWATERING. Postprocess cooling is necessary to achieve biosolids consolidation. A minimum of 20 days detention may be necessary for cooling and thickening. However, postcooling/thickening tanks sized for 1 to 3 days are adequate at some installations where heat exchangers are used to extract heat from the ATAD biosolids. Many German facilities use unmixed, unaerated open top concrete tanks with decant capability. Odor control typically is not provided and is not reported to be a problem. Post-ATAD biosolids typically gravity thickens to 6 to 10% solids (Deeney *et al.*, 1991), with some facilities reporting 14 to 18% solids by gravity. The Banff facility (5-day HRT) covered the postthickening facility for odor considerations. However, as a result, the biosolids cooled to 45 to 55°C during the summer, which resulted in high polymer consumption during dewatering. Ladysmith, British Columbia constructed a 10-day HRT covered storage tank and found dewaterability was best when biosolids from the ATAD were dewatered directly (Kelly *et al.*, 1993). Whistler uses a heat exchanger to cool digested biosolids to 35°C and heat incoming feed. At 35°C or less, dewatering is improved.

Limited data exist on post-ATAD biosolids dewaterability. However, the data available suggest the biosolids dewater similarly to anaerobically digested biosolids. Data from three Canadian facilities (Kelly *et al.*, 1993) are compared in Table 3.1 to typical data for dewatering for anaerobically digested biosolids on belt filter presses.

CONSTRUCTION FEATURES. In Germany, ATAD reactors typically are constructed of epoxy-coated cylindrical steel flat bottom tanks. The tank top and sides are insulated with 100 mm of mineral wood or sprayed-on polyurethane. The tank bottom is insulated with foam board. The insulation is protected with an aluminum or steel cladding. Access hatches are provided in the

Table 3.1 Comparison of autothermal thermophilic aerobic digestion processes (Kelly *et al.*, 1993, and WEF, 1992).

Biosolids type	Location	Dry solids feed, %	Dry polymer, g/kg dry solids	Cake solids, %
Autothermal thermophilic aerobic digestion	Whistler, BC[a]	5	3–8	25
Autothermal thermophilic aerobic digestion	Ladysmith, BC[b]	3–7	<1	12–15
Autothermal thermophilic aerobic digestion	Salmon Arm, BC[b]	3–5	25	30
Anaerobically digested	Typical	3–6	3–8	20–25

[a] Belt filter press.

[b] Screw press.

top of the reactor. The entire reactor is constructed above grade on a concrete foundation.

Typical height-to-depth ratios for aerators vary from 0.5 to 1.0. Actual ratios depend on the aerator type used to effect good mixing.

Heat exchange is not necessary for process requirements, but has been incorporated into some facilities for energy recovery and to preheat influent before the first stage reactor. According to literature, 50 to 70 MJ/m^3 can be recovered (Deeney *et al.*, 1991). Heat exchange can occur with heat exchangers during biosolids cooling or with a water cooling loop installed within the reactor shell. Also, heat from the offgas can be recovered. At Salmon Arm, British Columbia, heat recovery is 1.2 J/s and is used to heat the thickener building.

*P*ROCESS PERFORMANCE AND OPERATION

VOLATILE SOLIDS REDUCTION. Volatile solids reduction achievable by the process depends on the feed makeup, process HRT, operating temperature, and reactor loading.

Volatile solids reduction of 40% or greater has been reported in Germany for systems with detention times greater than 4 days. The Haltwhistle, United Kingdom, plant reported VSS destruction of 30 to 40% over a 2-year period. Canadian research found total volatile solids (TVS) destruction to be greater than 40% in a two stage reactor with a degree day product of 400°C days or greater. The Whistler, British Columbia, ATAD system has achieved TVS destruction of 70% (Kelly *et al.*, 1993).

PATHOGEN REDUCTION. The ATAD process in Germany is required to meet a limitation of 1 000 enterobacteria/mL in the product as mandated by regulations in Germany. This limitation is consistent with FRG regulatory requirements effective in 1987. The ATAD process in the FRG now maintains a status of approval by the FRG as a process that is capable of producing a "pasteurized (hygienic) sludge." This status is similar to the designation in the U.S. as a process to further reduce pathogens (PFRP) or a 40 CFR Part 503 rule designation of Class A.

The biosolids from treatment facilities that use the ATAD process are sampled twice per year by the regional health district. These samples are analyzed for a number of parameters including coliform count. If the biosolids are determined to meet the limitation of 1 000 cfu/mL (in addition to meeting other organic and inorganic criteria), they are approved as being "acceptable" for agricultural use.

The Haltwhistle, United Kingdom, facility (Murray *et al.*, 1990) reported pathogen log reductions of greater than 4. Canadian facilities (Kelly, 1991a) showed less than 100 MPN/wet gram for fecal coliform and fecal streptococci in 7 of 12 samples. Salmonella was not detected in any samples. Fecal streptococci was more than or near 100 MPN/wet gram in 5 of 12 samples. Tests by Jewell and Kabrick (1980) at Binghamton, New York, showed the ATAD to completely inactivate salmonella and viruses below limits of detection at 45°C and 24-hour detention time. U.S. EPA has indicated that the ATAD process should qualify for PFRP if the time–temperature requirements for Class A pathogen reduction are met.

The ATAD batch feed operating strategy uses the pathogen destruction potential of the process and reduces the chance for contamination. To prevent contamination with raw incoming feed, a specific volume of biosolids is removed on a daily basis from the second stage reactor (which is operating in a range of 55 to 65°C). After the biosolids are removed, biomass from the first stage reactor is transferred into it as a batch. The second stage is then not disturbed until the next batch is loaded 24 hours later. With this operating method, biomass that has been transferred from the first stage reactor is maintained at a thermophilic temperature for a minimum period of 24 hours. Raw feed is then introduced to the first stage to make up the volume removed. This feeding approach isolates the reactors from each other and reduces the potential for contamination of the product.

ODOR CONTROL. Of the six facilities in Germany toured by Deeney *et al.* (1991), none experienced offensive odors. Discharges from the ATAD reactors could be characterized as "musty" and similar to odors experienced from conventional aerobic digesters. Three facilities reported occasional odors. One of these facilities reported occasional odors when the reactor temperature exceeds 65°C, while the other two experienced some odor only during system feeding. Two of the facilities in close proximity to residential areas are equipped with exhaust gas scrubbers for the ATAD reactors. The application of these scrubbers is considered to be somewhat experimental. Performance is reported to be good; however, occasional odors have been noted with the scrubbers in place.

The Banff, Alberta, facility has a water scrubber on the ATAD exhaust. The dewatered biosolids there exhibited no odors and appeared well stabilized. The Haltwhistle plant experienced no odor complaints. The Salmon Arm, British Columbia, facility sends exhaust gases to a trickling filter with no reported odor problems. Two others, Ladysmith, British Columbia, and Gibsons, British Columbia, discharge ATAD exhaust to biological filters.

The Glenbard pure oxygen ATAD system experienced odors of a rotten broccoli nature during operation. Testing of offgases found the presence of dimethylsulfide, which is an indicator of anaerobic conditions. Offgas testing at Salmon Arm and Whistler (Kelly *et al.*, 1993) showed the presence of hydrogen sulfide, methyl disulfide, dimethylsulfide, ammonia, and unidentified organic compounds.

Typically, odors can be reduced and controlled if proper operating temperatures are achieved and the system is adequately mixed and aerated. Odor control measures should be provided based on the proximity of facilities to residences and the general public. Odor control measures typically used include water scrubbers, biofilters, compost/soil filters, and diversion of off-gases to other processes (trickling filter or activated sludge).

OPERATION AND MAINTENANCE CONSIDERATIONS. Process control consists of the following elements:

- Continuous recording/monitoring of process temperatures,
- Measuring feed temperatures weekly,
- Measuring pH of the treated biosolids weekly (pH > 8 is desirable), and
- Measuring the dry solids and volatile solids content of the feed and treated biosolids twice weekly.

German facilities typically report less than 1 h/d to operate and maintain the system. This implies that the process constraints are considered in the design of the system. Other operating parameters that have been monitored at Canadian facilities include oxygen reduction potential and volatile fatty acid (VFA). Tests at Banff found that well-stabilized biosolids had VFAs of between 200 and 700 mg/L as acetic acid, as compared with influent values of 2 200 to 3 500 mg/L as acetic acid (Fries, 1992).

Maintenance of the system depends to a large degree on the type (composition) of the feed, pretreatment provided, and feed characteristics. Facilities with proper screening and grit removal reduce the potential for rags and grit buildup and aerator abrasion from grit. Aspirating aerators may periodically plug with grease in the hollow shafts. Periodic jogging of the aerators on and off, and flushing the shaft may remove a buildup of grease.

*A**BILITY TO MEET U.S. ENVIRONMENTAL PROTECTION AGENCY REGULATIONS*

VECTOR ATTRACTION. Vector attraction reduction requirements relate to volatile suspended solids reduction (minimum of 38%) or specific oxygen uptake rate (SOUR) of less than 1.5 mg O_2/g·h total solids (dry weight basis) at a temperature of 20°C. Limited data are available on the ability of the process to meet these requirements. Data show VSS reduction ranging from

29 to 70%. This wide range suggests that VSS reduction is influenced by ATAD design differences, operating conditions, and feed characteristics. Therefore, the ability of the process to meet these conditions depends on a properly designed and operated system.

CASE HISTORIES

Tables 3.2 and 3.3 summarize available design or operation data on German two-stage facilities and Canadian facilities. Data were obtained through contact with system manufacturers, engineers, owners, operators, researchers, site visits, and published sources. Table 3.2 summarizes selected design data for 35 existing ATAD plants in Germany (Deeney *et al.*, 1991). Table 3.3 summarizes selected operations data for three Canadian facilities (Kelly *et al.*, 1993).

REFERENCES

Andrews, J.F., Kambhu, K. (1970) Thermophilic Aerobic Digestion of Organic Solid Wastes. Progress Report, Clemson Univ., S.C.

Deeney, K., *et al.* (1991) Autoheated Thermophilic Aerobic Digestion. *Water Environ. Technol.,* **65.**

Fries, M.K. (1992) Aerobic Digestion of Wastewater Sludge at Banff WWTF. Reid Crowther.

Gay, *et al.* (1974) High Purity Oxygen Aerobic Digestion Experiences at Speedway, Indiana. *Proc. Natl. Conf. Munic. Sludge Manage.,* Pittsburgh, Pa.

Jewell, W.J., and Kabrick, R.M. (1980) Autoheated aerobic thermophilic digestion with air aeration. *J. Water Pollut. Control Fed.,* **52,** 512.

Kelly, H.G. (1991a) Autothermal Thermophilic Aerobic Digestion: A Two Year Appraisal of Canadian Facilities. *Proc. Environ. Eng. Proc., Spec. Conf. Am. Soc. Civ. Eng.,* Reno, Nev.

Kelly, H.G. (1991b) Autothermal Thermophilic Aerobic Digestion of Municipal Sludges: Conclusions of a One-Year Full-Scale Demonstration Project. Paper presented at 64th Annu. Conf., Water Pollut. Control Fed., Toronto, Can.

Kelly, H.G., *et al.* (1993) Autothermal thermophilic aerobic digestion of municipal sludges: a one-year full scale demonstration project, *Water Environ. Res.,* **65,** 849.

Matsch, L.C., and Drnevich, R.F. (1977) Autothermal Aerobic Digestion. *J. Water Pollut. Control Fed.,* **49,** 296.

Murray, K.C., *et al.* (1990) Thermophilic Aerobic Digestion—A Reliable and Effective Process for Sludge Treatment at Small Works. *Water Sci. Technol.,* **22,** 225.

Popel, F. (1971a) Energieerzeugung beim biologischen Abbau Organischer Stoffe. *Gewaesser Abwaesser* (Ger.), **112.**

Popel, F. (1971b) Die theoretischen und praktischen Grunlagen der Flussigkompostierung Hockkonzentrierten Substrate. *Ausgearbeitet fur die Badische Anilin-und Sodafabrik Ludwigshafen, Stuttgart,* November.

Water Environment Federation (1992) *Design of Municipal Wastewater Treatment Plants.* Manual of Practice No. 8; Alexandria, Va.; Manual and Report on Engineering Practice No. 76; Am. Soc. Civ. Eng., New York, N.Y.

Wolinski, W.K. (1985) Aerobic Thermophilic Sludge Stabilization Using Air. *Water Pollut. Control,* 433.

Table 3.2   Summary of autothermal thermophilic aerobic digestion installations and design conditions in Germany.[a]

Name	Size, PE	Design loading, avg kg/d[b]	Feed volume, m/d	Number of reactors	Dimensions: diam, height, m^2	Active volume, ea, m^2	Solids retention time, d	Volumetric loading, kg/m^2·d	Number of aerators/ reactor	Installed aeration power, kW	Number of foam cutters/ reactor	Installed foam cutters/ reactor	Total installed power, kW	Power density, W/m^2	Daily power use, kW/d	Specific power applied, W/kg/VSS	Years in service
Backnang	10 000	800	20	2	3.3	3.3	3.3	12.1	2	11	2	1.5	12.5	180	300	375	0.75
Bad Grund	25 000[c]	2 000	50	—	—	—	—	6.7	3	27	3	4	31	120	744	433	3
Bergen	20 000	1 720	43	2	7.0,4.0	129	6.0	5.7	2	13.6	2	3	16.6	119	368	490	6
Bruderbach	8 500	800	20	2	—	70	7.0	9.7	2	20.4	2	4.5	24.9	208	598	515	3
Elwangen	22 000	1 160	29	3	5.0,4.0	60	4.1	7.33	18	2	2	20	83	480	273	7	—
Fassberg	22 000	1 760	40	2	7.0,3.2	120	5.5	6.7	1	5	2	0.6	5.6	117	134	420	9
Gemmingen	4 000	320	8	2	3.5,2.5	24	6.0	—	—	—	—	—	—	—	—	—	7
Isenbüttel	10 000	800	20	—	—	—	—	—	—	—	—	—	—	—	—	—	3
Jork	36 000	2 680	67	—	—	—	—	8.9	2	13.6	3	3	16.6	132	306	356	2
Kirchberg	12 000	1 120	28	2	5.0,4.0	63	4.5	8.3	2	22	—	—	—	—	—	—	5
Kronach	7 600	906	22.7	2	—	72	6.3	—	—	—	—	—	—	—	—	—	5
Lauterbach	80 000	4 800	120	—	6.5,4.0	—	—	6.7	3	23.6	2	3	26.6	127	638	458	2.5
Lengende[d]	17 000	1 400	35	2	6.5,4.0	105	6.0	—	—	—	—	—	—	—	—	—	1.4
Loshiem	12 000	520	13	—	—	—	—	6.8	1	5.75	2	1.5	7.25	145	174	512	0
Meinersen/Leiferde	4 000	340	8.5	2	3.5,3.5	25	5.9	—	—	—	—	—	—	—	—	—	0
Nesselwang	10 000	1 000	25	2	—	75	6.0	6.7	3	18.4	2	3	19.4	129	406	406	4
Nette	80 000	3 040	76	2	—	—	1 804.7	8.4	3	39	4	4.5	43.5	121	1 044	343	6
Neuhaus	12 500	1 000	25	—	—	—	—	—	—	—	—	—	—	—	—	—	4
Rangendingen	6 000	600	15	2	—	30	4.0	10.0	2	12	—	—	—	—	—	—	1
Rheinhausen	5 300	320	8	—	—	—	—	—	—	—	—	—	—	—	—	—	6
Römersberg	18 000	840	21	2	—	48+24	3.4	11.7	1	16	—	—	—	—	—	—	5

Table 3.2　Summary of authothermal thermophilic aerobic digestion installations and design conditions in Germany[a] (continued).

Name	Size, PE	Design loading, avg kg/d[b]	Feed volume, m/d	Number of reactors	Dimensions: diam, height, m^2	Active volume, ea, m^2	Solids retention time, d	Volumetric loading, $kg/m^2 \cdot d$	Number of aerators/ reactor	Installed aeration power, kW	Number of foam cutters/ reactor	Installed foam cutters/ reactor	Total installed power, kW	Power density, W/m^2	Daily power use, kW/d	Specific power applied, W/kg/VSS	Years in service
Schöntal	10 500	840	21	1	—	—	—	—	—	—	—	—	—	—	—	—	11
Schwarmstedt	20 100	1 440	36	2	7.0,4.0	115	—	6.3	3	24.8	3	4.5	29.3	127	703	408	2
St. Goarshausen	11 000	880	22	—	—	—	—	—	—	—	—	—	—	—	—	—	0
Thailen	7 000	820	8	—	—	—	—	—	—	—	—	—	—	—	—	—	7
Toging	15 000	1 120	28	—	—	—	—	—	—	—	—	—	—	—	—	—	4
Tönisberg	13 000	600	17	2	—	—	607.1	5.7	2	13.2	2	3	16.2	135	300	572	3
Trebur	8 000	640	16	—	—	—	—	—	—	—	—	—	—	—	—	—	0.25
Vilsbiburg	28 000	800	20	—	2	—	—	—	—	—	—	—	—	—	—	—	12
Walheim[d]	5 000	400	10	2	—	40	8.0	5.0	2	7.8	2	1.5	9.3	116	223	558	0.25
Wesendorf[d]	13 000	620	15.5	2	—	50	6.5	8.2	2	13	2	1.5	14.5	145	348	561	1
Winsen[d]	15 000	1 400	35	2	—	110	8.3	6.4	3	25	3	4.5	29.5	134	706	508	1

[a]　Fall 1989.
[b]　Design loading calculated at average 4% volatile suspended solids concentration.
[c]　Calculated value.
[d]　Equipped with two-speed aerators; power calculated at maximum speed.

Table 3.3   Demonstration facility: average operating criteria.*

Parameter	Ladysmith	Gibsons	Salmon Arm
Equivalent population	5 000	2 800	6 000
Plant flow, m^3/d			
Average sludge flow	2 000	1 150	2 300
Sludge	PS	PS+TF–SC	PS+FGR–SGR
Average solids production, kg/d	90–180	172	600
Average underflow concentration, %	5.4	4.1	4.6
Average underflow rate, m^3/d	1.5–3.0	4.2	13
Solids per capita, g/cap·d	20–36	62	100
Digesters (series)	2 circular	2 circular	2 rectangular
Diameter or width, m	3.5	2.29	2.59
Length, m	—	—	3.96
Height, m	3.0	2.74	4.26
SWD, m	1.9	2.49	3.35
Liquid volume each, m^3	18	10.1	33.6 average
Total volume organic loading			
TVS, kg/m^2·d	5–10	17	20
Air flow, m^3/m^3·h	0.46–0.94	0.8–1.0	0.5–1.5
Power density, kW/m^3	135–200	676	120–200
Total detention, hydraulic retention time			
or solids retention time days (average)	20	4.8	6.1

*PS =   primary sedimentation; TF–SC = trickling filter–solids contact; and FGR–SGR = enhanced biological phosphorus removal, fixed growth reactor–suspended growth reactor combination.

Chapter 4
Anaerobic Digestion

The purposes of anaerobic digestion are to produce stabilized biosolids, reduce pathogens, reduce biomass quantity by partial destruction of volatile solids, and produce usable gas as a byproduct.

BASIC THEORY

MICROBIOLOGY. Anaerobic digestion is the solubilization and reduction of complex organic substances by microorganisms in the absence of oxygen. The products of anaerobic digestion are methane, carbon dioxide, trace gases, and stabilized biosolids. The microbial population responsible for this conversion process can be divided into three groups, each responsible for a separate function: solubilization, acid formation, and methane formation. Figure 4.1 illustrates the anaerobic digestion process. Proteins, lipids, carbohydrates, and other complex organics are solubilized by hydrolysis. These products are then converted to short-chain organic acids including acetic, propionic, and lactic acid. The first two steps together are sometimes referred to as the acid-forming phase. These acids are then converted to methane, carbon dioxide, and other trace gases by methanogens (methane-forming bacteria).

The acid formers and methanogens perform their functions and produce acids and methane in sequence. The acid formers are tolerant to environmental changes such as pH and temperature. Their growth rate is relatively fast compared with methanogens. In contrast, the methanogens are sensitive to temperature changes, pH, and substrate composition. Their slow growth combined with their intolerance to changing environmental conditions can result in process upsets. For this reason, it is important to continuously monitor the existing conditions in the digester. If methane production is hindered, organic acids accumulate and cause the pH to drop. After pH starts dropping, methane production is further curtailed and the ratio of acid produced to methane produced increases. This imbalance may continue until the digester fails and stabilization stops. Anaerobic digesters are designed on the basis of methane formation being the limiting step in the process.

Studies by Eastman and Ferguson (1981) indicate that for complex particulate substrates, such as wastewater residuals, the hydrolysis is the rate-limiting step for the acid-forming phase. After these complex organic substances are solubilized, they are immediately converted to volatile acids and do not accumulate.

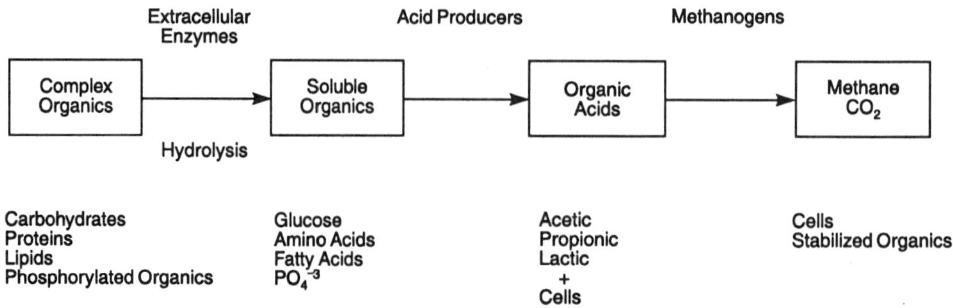

Figure 4.1   Microbiological pathway of anaerobic digestion.

PROCESS DESCRIPTION. Low-Rate and High-Rate Digestion. Typically, anaerobic digesters are classified into low-rate and high-rate digesters. Low-rate digesters are the oldest anaerobic stabilization systems and are also called standard-rate or conventional anaerobic digesters. Figure 4.2 illustrates the low-rate digester. The digester consists of a cylindrically shaped tank with a sloping bottom and a flat or domed roof. No mixing is provided in this system.

Because no mixing occurs, stabilization in low-rate systems results in a stratified condition within the digester. Methane gas accumulates in the headspace of the tank and is drawn off for storage or use. Scum accumulates on the liquid or supernatant surface. The supernatant is drawn off and recycled either to the primary clarifier or to the secondary treatment process. The supernatant typically contains high ammonia and phosphorus concentrations. The stabilized biosolids settle to the tank bottom for removal and further processing.

Low-rate digestion is characterized by intermittent feeding, low organic loading rates, no mixing other than that caused by rising gas bubbles, large tank size because of the small effective volume, and detention times of 30 to

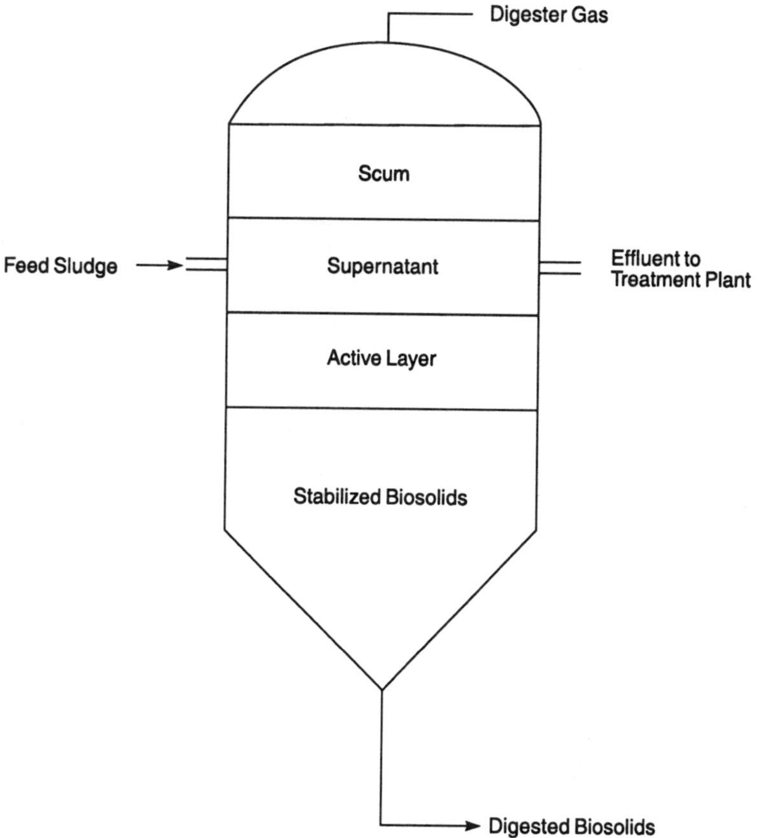

Figure 4.2   Low-rate anaerobic digestion (U.S. EPA, 1979).

---

60 days. Grit and scum layers will accumulate on the bottom and top of the tank, respectively, decreasing the effective volume. An external heat source may or may not be present to increase the digestion rate. Typically, conditions are not maintained for optimum digestion. This type of digestion has traditionally been considered only for small wastewater treatment plants (WWTPs) (under 40 L/s [1.0 mgd]); however, their use has decreased in recent years.

Several improvements to the low- or standard-rate system were developed in the 1950s, resulting in the high-rate anaerobic digestion system. This high-rate system is characterized by heating, auxiliary mixing, thickening, and uniform feeding (see Figure 4.3). An attempt is made to maintain ideal environmental conditions so the microorganisms will perform according to laboratory studies and theory.

HEATING. Heating during digestion increases the microorganism growth rate, the digestion rate, and gas production. High-rate anaerobic digesters are operated at mesophilic and thermophilic temperature ranges. The mesophilic range is approximately 30 to 38°C (85 to 100°F) and the thermophilic range is 50 to 60°C (122 to 140°F). Temperature variations have a negative effect on the methanogens, which can cause process upsets. In this regard, thermophilic microorganisms are considered to be more sensitive to temperature changes than mesophilic organisms.

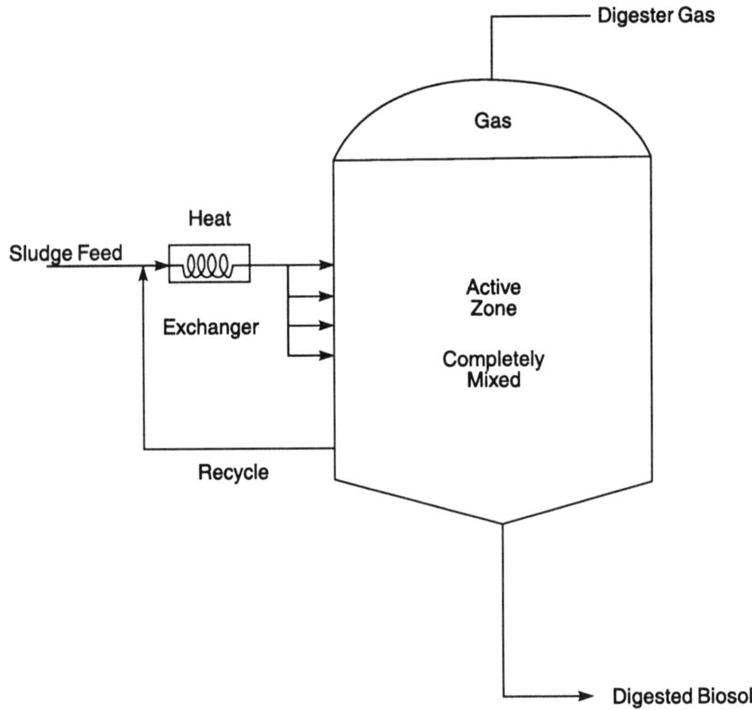

Figure 4.3   High-rate anaerobic digestion.

Thermophilic digestion may offer several advantages over mesophilic digestion, including increased reaction rates that can result in smaller digester volumes, increased destruction of pathogens, and better dewatering characteristics. Several large cities in the U.S., including Los Angeles and New York, have used the thermophilic process with limited success. Limitations of the process include extreme sensitivities of the organisms to the defined temperature range, a higher net energy input (therefore higher operation costs) than that for the mesophilic, and the production of biosolids with a comparatively more offensive odor.

Several heating methods have been used for anaerobic digesters, including steam injection, internal heat exchangers, and external heat exchangers. External heat exchangers are the most popular because of their flexibility and ease of maintaining the heating surfaces. Internal coils may foul from caking and have to be removed or the digester has to be emptied to clean them. Hot water supply temperatures to heat exchanger surfaces are maintained between approximately 50 and 62°C (120 and 150°F). At temperatures higher than 62°C (150°F), caking on the heat exchanger surfaces becomes more likely. The use of steam injection results in dilution. A considerable amount of water is necessary for steam production. Costs for raw water and softening add to the cost of steam injection.

MIXING. Auxiliary mixing of the digester contents is beneficial for the following:

- Reducing thermal stratification,
- Dispersing the substrate for better contact with the active biomass,
- Reducing scum buildup,
- Diluting any inhibitory substances or adverse pH and temperature feed characteristics,
- Increasing the effective volume of the reactor,
- Allowing reaction product gases to separate more easily, and
- Keeping in suspension more inorganic material that has a tendency to settle.

Three types of mixing methods have typically been used: mechanical, pumped, and gas recirculation. Mechanical or internal mixers use impellers, propellers, and turbine wheels to mix. Problems arise because of the exposed surfaces of the mixers. Shafts and impellers are affected by vibration because of collected materials and wear resulting from grit and debris. Mixing by pumped recirculation involves recycling the digester contents with an external pump. The efficiency of this type of mixing depends on the digester size, net energy input, viscosity, and turnover rate.

Gas recirculation systems can be further classified as either unconfined or confined gas injection systems. Unconfined systems include top-mounted lances and diffusers located on the tank bottom. Confined systems discharge the gas through draft tubes. Gas recirculation systems include the use of tubes, sequentially operated lances, diffusers located on the tank bottom, and unconfined gas bubble release using a 0.3-m (1-ft) diam tube. In each case the digester gas produced is compressed and recirculated through the tank to promote mixing. Each of these systems has its advantages and disadvantages and the degree of mixing typically depends on the energy input.

FEEDING. Continuous or regular intermittent feeding is beneficial to digester operation because it helps maintain steady-state conditions within the digester. The methanogens are sensitive to changes in substrate levels. Uniform feeding and multiple feed point locations in the tank can alleviate or reduce shock loading to those microorganisms. Excessive hydraulic loading, which decreases detention time, dilutes alkalinity necessary for buffering capacity, and requires additional heat energy, should be avoided.

Two-Stage Anaerobic Digestion. Two-stage digestion is an expansion of the high-rate digestion technology that divides the functions of fermentation and solids–liquid separation into two separate tanks in series (Figure 4.4). The first tank is a high-rate stabilization system, while the second is for solid–liquid phase separation. The second reactor typically does not have mixing or heating facilities unless it is also used to provide standby digester capacity. It may serve several other functions such as providing storage capacity and insurance against short circuiting of the process.

Anaerobically digested biosolids may not stratify well, resulting in a supernatant containing a high concentration of suspended solids that can be detrimental to the liquid wastewater treatment system when recirculated. Several reasons for poor settling characteristics include incomplete digestion in the

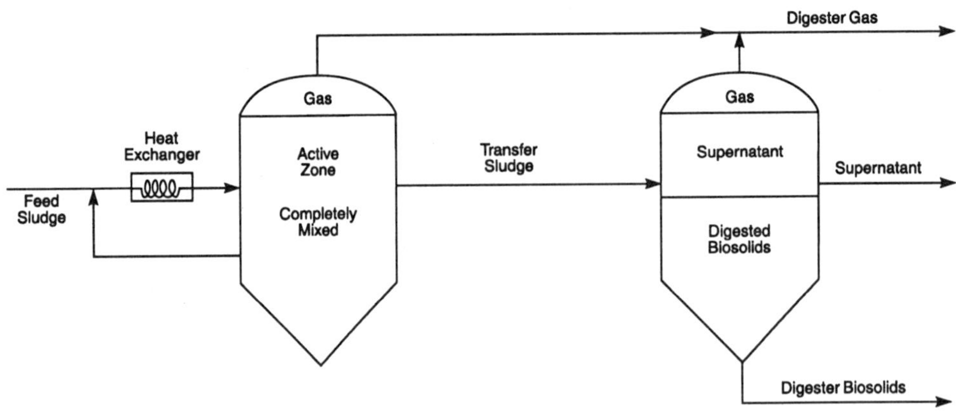

Figure 4.4   Two-stage, high-rate anaerobic digestion system (U.S. EPA, 1979).

primary digester (which generates gases in the secondary digester and causes floating solids) and fine-size solids that have poor settling characteristics. This latter case is associated with secondary or tertiary solids, including chemical solids. Primary- and waste-activated sludges thickened to more than 4 to 5% total solids typically will not separate in the second digester.

STABILIZATION CRITERIA. Volatile Solids. The anaerobic digestion process is used to stabilize and reduce the quantity of volatile solids for disposal. The degree of volatile solids destruction is one indicator for the operator to determine if the digester is functioning properly. A typical high-rate digester will reduce the volatile solids content by 40 to 60%. Each digester will display its own maximum level of destruction depending on feed characteristics. Various investigations indicate that 60 to 80% of the measured volatile solids fed to the digester are biodegradable, unless industrial wastes or other abnormal characteristics are present.

The amount of volatile solids destroyed is a function of both temperature and solids retention time (SRT). Investigations by O'Rourke (1968) demonstrate the effects of temperature and SRT on volatile solids destruction. Figure 4.5 illustrates these effects on a laboratory-scale study of mesophilic digestion.

Typically, increasing the SRT will increase the percent volatile solids destruction to 50 to 60% at 35°C. After an optimum SRT is reached, additional destruction is minimal. At different temperatures, the total amount of volatile solids destroyed increases for increasing temperatures up to 35°C or the mesophilic range. Literature indicates that primary solids degrade faster than

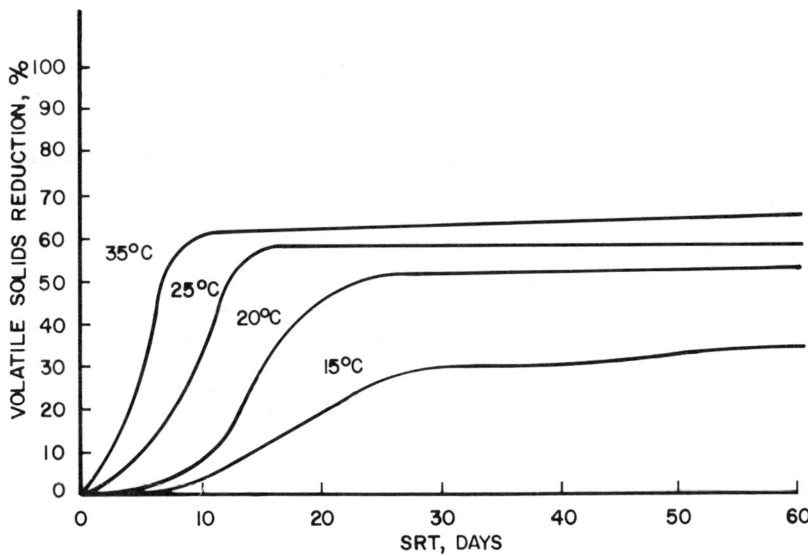

Figure 4.5 **Volatile solids destruction as a function of temperature and solids retention time (O'Rourke, 1968).**

activated-sludge solids, and a mixture of both degrades at a rate in between that of each separately. It is also important to note that the overall volatile destruction potential could vary as a function of the liquid treatment process that generates it. Typically, primary solids from domestic wastewater should have approximately equal biodegradable volatile fractions. Secondary solids, however, will vary depending on the operating conditions of the liquid process. For example, when comparing extended aeration and high-rate processes, the biodegradable fraction of the volatile solids of the waste-activated sludge generated would be lower for the extended aeration process, given the effect of endogenation. Therefore, under identical conditions, the digested biosolids from the high-rate system would show a greater volatile solids destruction (hence higher gas production). However, both products would have reached a similar level of stabilization.

Haug *et al.* (1978) have published data to demonstrate that by thermally treating activated sludge for 0.5 hours at 175°C (345°F) before digestion, solids reduction, gas production, and dewatering characteristics are improved. Woods and Malina (1965) have analyzed the results of digesting combined feed. Table 4.1 shows the results of high-rate, two-stage digestion operating in the mesophilic range. The second stage was used only for solids–liquid separation with no heating or mixing.

Pathogen Reduction. Anaerobic stabilization also accomplishes substantial reductions in pathogen concentration (Farell *et al.*, 1986; Garber, 1982; Kun *et al.*, 1989; Storey, 1987; Stukenberg *et al.*, 1992; and U.S. EPA, 1979). Viruses,

Table 4.1   Average physical and chemical characteristics of biosolids from two-stage digesters.

	Concentration, mg/L*			
Component	**Feed sludge**	**Transfer sludge**	**Supernatant**	**Stabilized biosolids**
pH	5.7	7.7	7.8	7.8
Alkalinity	758	2 318	2 630	2 760
Volatile acids	1 285	172	211	185
Total solids	35 600	18 200	12 100	32 800
Fixed solids	9 000	6 600	3 310	12 300
Carbohydrates	9 680	1 550	1 020	3 100
Lipids	8 310	2 075	1 321	3 490
Carbon	15 450	6 950	4 440	10 910
Proteins, as gelatin	18 280	11 200	6 580	17 200
Ammonia nitrogen, as NH_3	213	546	618	691
Organic nitrogen, as NH_3	1 346	879	564	1 455
Total nitrogen, as NH_3	1 559	1 425	1 182	2 146

* Except pH.

bacteria, and parasites entering WWTPs will eventually accumulate in the biosolids generated. Pathogen reduction is required to reduce public health concerns related to biosolids handling and beneficial use. The U.S. Environmental Protection Agency (U.S. EPA) has established regulations that dictate specific pathogen reduction performance requirements for the beneficial use of anaerobically digested biosolids (see Appendix). Table 4.2 summarizes the effect of the digestion process on pathogen concentrations. Anaerobic digestion has been shown to reduce detectable viruses by one to four orders of magnitude, with the higher reductions achieved at thermophilic operating ranges. Reductions of two orders of magnitude of fecal coliforms typically are observed in well-operated systems. With regard to parasites, protozoa cysts typically do not survive anaerobic digestion. However, some helminth ova survive even after thermophilic digestion. Given that complete pathogen elimination through anaerobic digestion is not feasible, further restrictions are imposed limiting human contact with the biosolids.

Gas Quality. The quality and quantity of digester gas produced can also be used to evaluate digester performance. Gas production is directly related biochemically to the amount of volatile solids destroyed and is expressed as volume of gas per unit mass of volatile solids destroyed. This specific gas production rate is different for each specific organic substance in the digester.

Table 4.2 Typical pathogen concentrations (Farell *et al.*, 1986; Ghosh *et al.*, 1991; Kun *et al.*, 1989; Stukenberg *et al.*, 1992; and U.S. EPA, 1979).

	Concentration, number/100 mL	
Pathogen	**Unstabilized raw solids[a]**	**Digested biosolids[b]**
Virus (various)	$380 - 7{\times}10^4$	$BDL^c{-}10^3$
Bacteria		
Total coliforms	$4.3{\times}10^9 - 5{\times}10^9$	$3{\times}10^4 - 7{\times}10^7$
Fecal coliforms	$1.4{\times}10^9 - 10^9$	$BDL - 7.8{\times}10^6$
Salmonella	$3 - 4.6{\times}10^4$	$3 - 62$
Streptococcus fecalis	$2.3{\times}10^7 - 1.5{\times}10^8$	$BDL - 2.2{\times}10^6$
Mycobacterium tuberculosis	10^7	10^6
Escherichia coli	$9.5{\times}10^7$	BDL
Parasites		
Ascaris	$200 - 10^4$	$0 - 10^3$
Helminth eggs	$20 - 700$	$30 - 70$
Tapeworm eggs	$2{\times}10^3$	2

[a] Type of solids typically unspecified.

[b] Anaerobic digestion; configuration, temperature, and detention times varied.

[c] BDL = Below detection limits.

Table 4.3   Typical gas production rates (Buswell and Neave, 1939).

Material	Specific gas production per unit mass destroyed	
	m^3/kg	Methane content, %
Fats	1.2–1.6	62–72
Scum	0.9–1.0	70–75
Grease	1.1	68
Crude fibers	0.8	45–50
Protein	0.7	73

Table 4.3 gives the specific gas production for several organic substances. The range of gas production varies from approximately 1.2 to 1.5 m^3/kg (20 to 25 cu ft/lb) volatile solids destroyed for fats to 0.7 m^3/kg (12 cu ft/lb) of volatile solids destroyed for proteins and carbohydrates. A typical municipal anaerobic digester handling primary and waste-activated sludge should produce approximately 0.8 to 1 m^3/kg (13 to 18 cu ft/lb) of volatile solids destroyed. The amount of gas produced is a function of temperature, SRT, and volatile solids loading. Specific gas production should be measured until an average value can be obtained and used for monitoring.

Methane and carbon dioxide are the two main constituents of digester gas, with trace amounts of nitrogen, hydrogen, and hydrogen sulfide. Expected performance data from healthy digesters suggest methane concentrations of 65 to 70% by volume and carbon dioxide concentrations of 30 to 35% by volume. Tortorici and Stahl (1977) have published data showing typical digester gas characteristics (see Table 4.4). The presence of hydrogen sulfide can indicate unbalanced digestion, industrial waste sources, or saltwater infiltration. Hydrogen sulfide may be responsible for odor problems and excessive corro-

Table 4.4   Digester gas characteristics (Tortorici and Stahl, 1977).

Constituent	Values for various plants, % by volume[a]							
Methane	42.5	61.0	62.0	67.0	70.0	73.7	75.0	73 – 75
Carbon dioxide	47.7	32.8	38.0	30.0	30.0	17.7	22.0	21 – 24
Hydrogen	1.7	3.3	[b]	—	—	2.1	0.2	1 – 2
Nitrogen	8.1	2.9	[b]	3.0	—	6.5	2.7	1 – 2
Hydrogen sulfide	—	—	0.15	—	0.01	0.06	0.1	1 – 1.5
Heat value, Btu/cu ft[c]	459	667	660	624	728	791	716	739 – 750
Specific gravity (air=1)	1.04	0.87	0.92	0.86	0.85	0.74	0.78	0.70 – 0.80

[a] Except as noted.

[b] Trace.

[c] Btu/cu ft $\times$ 37.26 = kJ/m^3.

sion in the digester and adjacent piping. Heavy metals may precipitate out as sulfide and reduce the hydrogen sulfide concentration in the digestion gas stream. An increase in carbon dioxide levels (%) is often an indication of an upset digester.

Typical digester gas exhibits a heat content of between 20 and 25 MJ/m^3 (500 and 700 Btu/cu ft). An average value of 25 MJ/m^3 (640 Btu/cu ft) has been used for design. The heat content in MJ/m^3 is approximately one-third the percent methane.

APPLICABILITY. Anaerobic digestion may be considered beneficial for stabilization when the volatile solids content is 50% or higher and if no inhibitory substances are present or expected. Digestion of primary solids results in better solids–liquid separation characteristics than activated sludge. Combining components will result in settling characteristics better than activated sludge but less desirable than primary alone. Chemical residuals containing lime, alum, iron, and other substances can be successfully digested if the volatile solids content remains high enough to support the biochemical reactions and no toxic compounds are present. If an examination of past sludge characteristics indicates wide variations in quality, anaerobic digestion may not be feasible because of its inherent sensitivity to changing substrate quality.

The advantages offered by anaerobic digestion include

- Excess energy (more than that required by the process) is produced. Methane is produced and can be used to heat and mix the reactor. Excess methane gas can be used to heat space or produce electricity, or as engine fuel. However, gas cleaning systems are often required to render the digester gas useful for these purposes.
- The quantity of total solids for disposal is reduced. The volatile solids present are converted to methane, carbon dioxide, and water, thereby reducing the quantity of solids. Approximately 30 to 40% of the total solids may be destroyed and 40 to 60% of the volatile solids may be destroyed. This is particularly important when considering disposal or beneficial use options involving long hauling distances.
- The product is stabilized biosolids that can be used for land application.
- Pathogens are destroyed to a high degree during the process. Thermophilic digestion enhances the degree of pathogen destruction.
- In municipal treatment facilities, most organic substances are readily digestible, except lignins, tannins, rubber, and plastics.

The disadvantages associated with anaerobic digestion include

- The digester is easily upset by unusual conditions and erratic or high loadings, and slow to recover.
- Requires comparatively higher operational control.

- Heating and mixing equipment are required for satisfactory performance, making this process comparatively more equipment-intensive.
- Large reactors are required because of the slow growth of methanogens and required SRTs of 15 to 20 days for a high-rate system. Therefore capital costs are high.
- During digestion, heavy metals are concentrated in the biosolids. This may have a negative regulatory effect depending on the disposal technique.
- The resultant supernatant sidestream is a strong waste stream that adds to the loading of the WWTP. It contains high concentrations of biochemical oxygen demand (BOD), carbonaceous oxygen demand (COD), suspended solids, phosphorus, and ammonia nitrogen.
- Cleaning operations are difficult because of the closed vessel. Internal heating and mixing equipment can become major problems as a result of corrosion and wear in harsh, inaccessible environments.
- Biosolids poor in dewatering characteristics are produced. Also the heating value is reduced for incineration purposes.
- The possibility of explosion as a result of inadequate operation and maintenance, leaks, or operator carelessness exists.
- Gas line condensation or clogging can cause maintenance problems.
- There are comparatively high maintenance requirements because of potential scaling, scum buildup, and grit accumulation.

*D*ESIGN CONSIDERATIONS

The first important determination a design engineer must make is the correct reactor volume to ensure sufficient stabilization of the feed and methane gas production. Other decisions, such as mixing method, heating requirements, and energy recovery, are discussed later. This discussion is oriented toward two-stage mesophilic digestion because it is common in the U.S. Additional design considerations would be required for thermophilic or two-phase digestion.

DESIGN DATA. The data necessary for designing an anaerobic digester include the quality and quantity of feed solids to be digested. This includes the quantity of solids (mass per day) produced by primary, secondary, and tertiary (where applicable) treatment processes. Additional information is necessary on the percent total solids, percent volatile solids, and the ratio of primary to secondary solids to be fed to the digester. The total solids produced can be calculated theoretically based on the solids balance (see Chapter 1) or extrapolated from existing WWTP operating data. The percent total solids and volatile solids can be estimated or analyzed by the laboratory. The amount of grit is a concern because of its ability to decrease the effective volume of the digester if allowed to accumulate.

Given the total mass of solids produced from primary and secondary treatment, the corresponding volume can be calculated using the following equation

$$V_s = \frac{W_s}{K' s_g f}$$ (4.1)

Where

V_s =   volume of sludge, L/d (gpd);
s_g =   specific gravity of sludge;
f =   mass fraction of solids;
W_s =   dry mass of solids produced, kg/d (lb/d); and
K' =   unit mass of water.

The volume in cubic meters per day can be calculated by converting litres to cubic meters

$$V, \, m^3/d = \frac{V_s}{0.001 \, L/m^3}$$ (4.2)

The specific gravity may be calculated from the mass fractions of the fixed and volatile solids as

$$\frac{1}{s_g} = \frac{W_w}{s_{gw}} + \frac{W_f}{s_{gf}} + \frac{W_v}{s_{gv}}$$ (4.3)

Where

s_{gw} =   specific gravity of water (1.0);
s_{gf} =   specific gravity of fixed solids (approximately 2.5);
s_{gv} =   specific gravity of volatile solids (approximately 1.3);
W_w =   mass fraction of water;
W_f =   mass fraction of fixed solids, and
W_v =   mass fraction of volatile solids.

Alternatively, the specific gravity can be estimated as

$$s_g = 10 + 0.005 \, TS$$ (4.4)

Where

TS   =   percent total solids.

DESIGN PARAMETERS. Digester Volume. Design of anaerobic digesters has been based on SRT, organic loading rate (volatile suspended solids [VSS] per volume), and volume per capita. Typical design parameters for low- and high-rate digesters are detailed in Table 4.5. For estimating feed volumes at strictly domestic plants in the absence of operating data, the volume figures per capita may be used. Low-rate digesters are typically organically loaded at rates approximately equal to 0.5 to 1.5 kg/m^3·d (0.04 to 0.1 lb VSS/d/cu ft).

High-rate digesters with mixing and heating are characterized by organic loading rates of 2 to 3 kg/m^3·d (0.1 to 0.2 lb/d/cu ft).

Typical SRTs are approximately 30 to 60 days for standard or low-rate digestion and 15 to 20 days for high-rate digestion at mesophilic temperatures. The SRT may be defined as the ratio of the total mass of solids in the system to the quantity of solids withdrawn per day. For anaerobic digesters with no internal recycle, the SRT equals the hydraulic retention time. In the case of recycling settled solids, the SRT would increase to more than the hydraulic retention time. This recycle feature is characteristic of the anaerobic contact or two-phase digestion process.

A minimum SRT is essential to the anaerobic digestion process to ensure that the necessary microorganisms are being produced at the same rate as they are wasted daily. This critical SRT is different for various constituent groups. The lipid-metabolizing bacteria are the slowest growing and, therefore, need a longer SRT, while the cellulose-metabolizing bacteria require a shorter SRT (see Figure 4.6).

As the SRT is reduced below a critical time, the system fails from washout of the microbial population of methanogens. Lawrence (1971) has published minimum values of SRT required for reduction of several specific substances. Table 4.6 lists these SRTs, which range from less than 1 day for hydrogen to 4.2 days for wastewater solids, as a function of temperature. An increase in temperature reduces the SRT necessary for maximum performance (Figure 4.7). A temperature increase results in an increase in specific gas production. An SRT

Table 4.5   Typical design parameters for low-rate and high-rate digesters (Burd, 1968).

Parameter	Low rate	High rate
Solids retention time, days	30–60	15–20
Volatile suspended solids loading, lb/cu ft/d (kg/m^3·d)	0.04–0.1 (0.64–1.6)	0.10–0.20 (1.6–3.2)
Volume criteria, cu ft/cap (m^3/cap)		
Primary sludge	2–3 (0.06–0.08)	1.3–2 (0.03–0.06)
Primary sludge + trickling filter sludge	4–5 (0.11–0.14)	2.6–3.3 (0.07–0.09)
Primary sludge + waste-activated sludge	4–6 (0.11–0.17)	2.6–4 (0.07–0.11)
Combined primary + waste biological sludge feed concentration, % solids-dry basis	2–4	4–6
Anticipated digester underflow concentration, % solids-dry basis	4–6	4–6

of at least 10 days typically is used for mesophilic high-rate digesters. However, for stability and ease of control and to account for grit and scum accumulation and imperfect mixing, most digesters operate at 15 days retention or
higher.

Benefield and Randall (1980) have presented the kinetic model of a completely mixed reactor without recycle. From this model the critical SRT may
be evaluated by

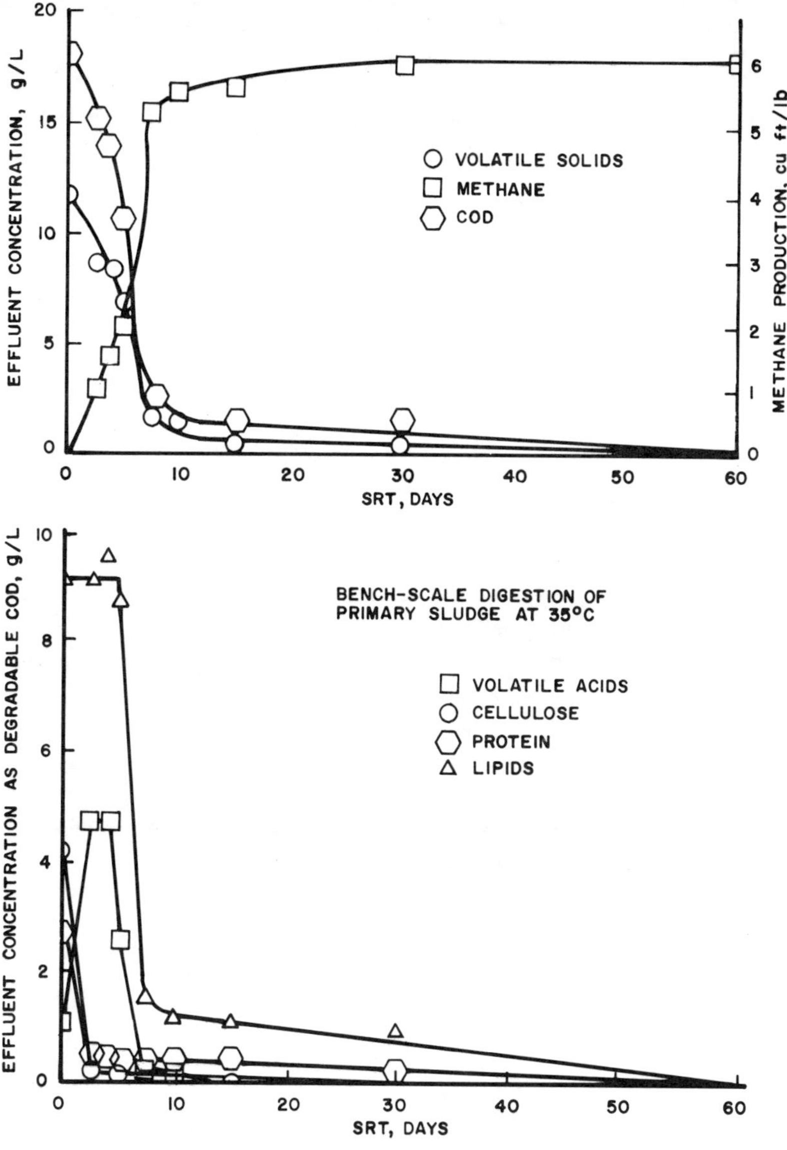

Figure 4.6   **Effect of solids retention time on the relative breakdown of
degradable waste components and methane production
(cu ft/lb × 0.062 43 = m³/kg).**

$$\frac{1}{\theta_c^m} = Y_t \frac{kS_o}{k_s + S_o} - K_d \tag{4.5}$$

Where

θ_c^m	=	critical SRT, day;
Y_t	=	yield coefficient;
k	=	maximum specific substrate utilization rate, day^{-1};
S_o	=	influent substrate concentration, mass per volume;
k_s	=	saturation constant, mass per volume; and
K_d	=	decay coefficient, time^{-1}.

Lawrence (1971) gives values (municipal sludge) for Y_t and K_d as 0.04 and 0.015 per day. O'Rourke (1968) gives values for k and K_s of 6.67 per day and 2 224 mg/L as COD for 35°C. The values of k and K_s should be corrected for temperatures below 35°C.

After the critical SRT is calculated or chosen from laboratory data, an appropriate safety factor (SF) is used to calculate the design SRT (θ_d^m), or

$$SRT = SF \times SRT \tag{4.6}$$

or

$$\theta_d^m = SF \times \theta_c^m \tag{4.7}$$

Lawrence and McCarty (1974) recommend an SF of 2 to 10 depending on expected peak loading and projected grit and scum accumulation in the tank. Smaller digesters should use higher SFs.

An alternative method of selecting a design SRT is to perform bench- or pilot-scale studies with representative feed and apply the appropriate kinetic equations for evaluating the constants. This technique may become necessary

Table 4.6   **Minimum values of solids retention time (θ_c) for anaerobic digestion of various substances (Lawrence, 1971).**

Substrate	35°C	30°C	25°C	20°C
Acetic acid	3.1	4.2	4.2	—
Propionic acid	3.2	—	2.8	—
Butyric acid	2.7	—	—	—
Long-chain fatty acid	4.0	—	5.8	7.2
Hydrogen	0.95[a]	—	—	—
Wastewater sludge	4.2[b]	—	7.5[b]	10

[a]   For 37°C.

[b]   Computed value.

when contributions from industrial wastes are significant. The effect of industrial wastes on anaerobic digestion is always questionable until the specific waste is characterized and tested.

After the selection of an appropriate design SRT, the digester volume may be found by using the expected daily flow or

$$V_R = + V_s \, \theta_d^m \qquad (4.8)$$

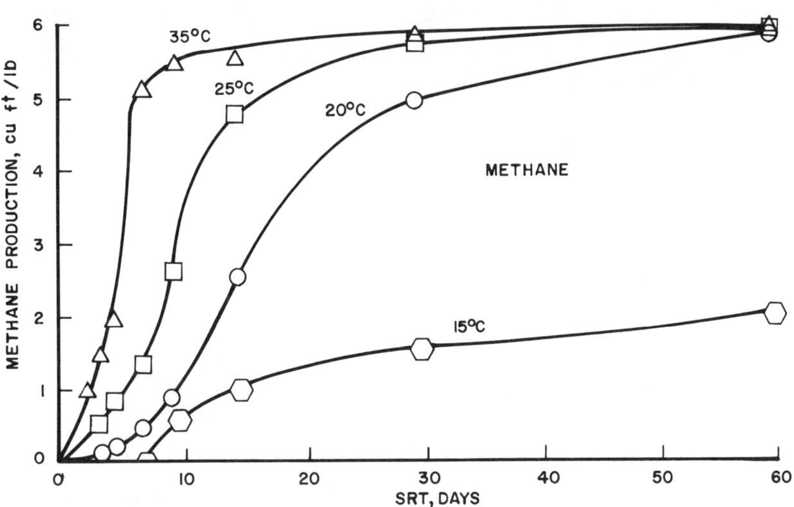

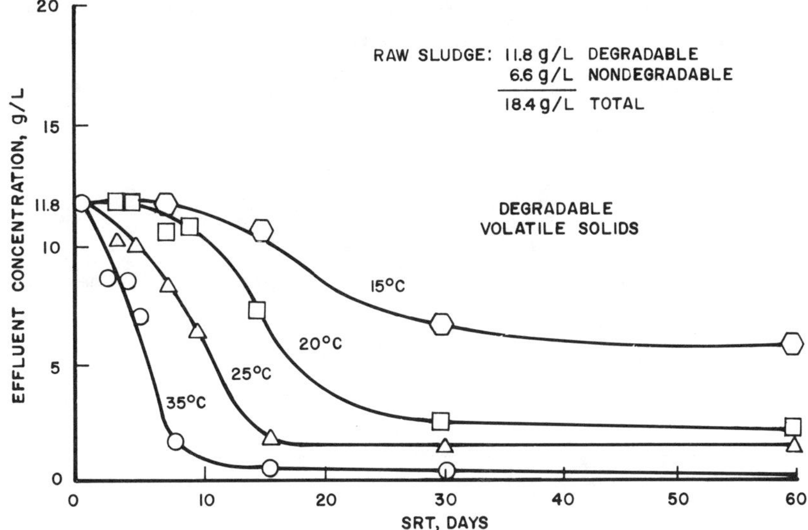

Figure 4.7 **Effect of temperature solids retention time on the pattern of methane production and volatile solids breakdown (O'Rourke, 1968).**

Where

$$V_R \quad = \quad \text{digester volume, m}^3 \text{ (cu ft), and}$$
$$V_s \quad = \quad \text{daily sludge rate, m}^3\text{/d (cu ft/d).}$$

For a circular digester, the sidewater depth may vary from 8 to 12 m (25 to 40 ft). The cross-sectional area and diameter can then be calculated.

Volatile Solids Destruction Estimation. The amount of volatile solids destruction expected can be estimated from previous data (40 to 60%) or equations relating volatile solids destruction to detention time (Liptak, 1974). For a standard rate system

$$V_d = 30 + \frac{t}{2} \tag{4.9}$$

Where

$$V_d \quad = \quad \text{volatile solids destruction, \% and}$$
$$t \quad = \quad \text{time of digestion, days.}$$

For a high-rate digestion system:

$$V_d = 13.7 \ln (\theta_d^m) + 18.94 \tag{4.10}$$

Additional estimates can be made from Table 4.7. The concentration of fixed solids entering the digester will remain constant.

A close estimation of the solids (kg/d or lb/d) that would enter the second-stage digester of a two-stage system is given by

$$\text{Solids} = \text{TS} - (A \times \text{TS} \times V_d) \tag{4.11}$$

Where

$$\text{TS} \quad = \quad \text{total solids entering primary digester, kg/d (lb/d);}$$
$$A \quad = \quad \text{volatile solids; \%, and}$$
$$V_d \quad = \quad \text{volatile solids destroyed in primary digester, percent.}$$

Table 4.7   Estimated volatile solids destruction.

	Digestion time, days	Volatile solids destruction, %
High rate (mesophilic range)	30	65.5
	20	60.0
	15	56.0
Low rate	40	50.0
	30	45.0
	20	40.0

Equation 4.8 can be used to size the second-stage digester after the total solids entering the secondary digester, the percentage the solids will thicken to, and the storage period required for disposal are known.

Gas Production and Quality. The specific gas production at municipal plants can be estimated by using the relationship of approximately 0.8 to 1.1 m^3/kg (13 to 18 cu ft/lb) of volatile solids destroyed. The greater the percentage of fats and grease, the higher the expected specific gas production as long as adequate SRT and mixing is provided, because these materials are the slowest to metabolize. The total gas volume produced equals

$$G_v = (G_{SGP}) \, V_s \qquad (4.12)$$

Where

$$
\begin{aligned}
G_v &= \text{volume of total gas produced, } m^3 \text{ (cu ft);} \\
V_s &= \text{volatile solids destroyed, kg (lb); and} \\
G_{SGP} &= \text{specific gas production, taken as 0.8 to 1.1 } m^3\text{/kg (13 to} \\
&\quad\ \text{18 cu ft/lb) VSS destroyed.}
\end{aligned}
$$

The total amount of methane produced can be estimated from the amount of organic material removed each day by the following relationship:

$$G_m = M_{SGP} \, (\Delta OR - 1.42 \, [\Delta X]) \qquad (4.13a)$$

or

$$G_m = \frac{(\Delta COD_R)(13.1 \text{ MJ/kg COD}_R)}{38 \text{ MG/m}^3 \text{ CH}_4} \qquad (4.13b)$$

Where

$$
\begin{aligned}
G_m &= \text{volume of methane produced, } m^3\text{/d (cu ft/d);} \\
M_{SGP} &= \text{specific methane production per mass of organic} \\
&\quad\ \text{material (BOD or COD) removed, } m^3\text{/kg (cu ft/lb);} \\
\Delta OR &= \text{organics (COD or BOD) removed daily, kg/d (lb/d);} \\
\Delta X &= \text{biomass produced kg/d (lb/d); and} \\
\Delta COD_R &= \text{COD removed across digester.}
\end{aligned}
$$

Because digester gas is approximately two-thirds methane, the total digester gas produced equals

$$G_T = \frac{G_m}{0.67} \qquad (4.14)$$

Where

$$G_T = \text{total gas produced, } m^3\text{/d (cu ft/d).}$$

Expected methane concentrations for different digesters may range from 45 to 75% in volume (see Table 4.4). Carbon dioxide concentrations will range between 25 and 45% by volume. The presence of hydrogen sulfide should be investigated by determining any sources of industrial waste or salt water infiltration. The expected heat value of digester gas is approximately 24 MJ/m^3 (640 Btu/cu ft), compared to 38 MJ/m^3 (1010 Btu/cu ft) for methane.

PROCESS REQUIREMENTS. Mixing Requirements. Most manufacturers of digester mixing equipment will suggest the appropriate type, size, and power level of mixing equipment depending on the digester volume and geometry. These suggestions typically are based on in-house studies and past successful experience of other similar installations. Anaerobic digesters can be mixed with gas, mechanical, or pumped mixing systems (various mixing systems have different advantages and disadvantages). Selection of a mixing system is based on costs; maintenance requirements; process configuration; and the screenings, grit, and scum content of the feed. Suggested parameters for sizing digester mixing systems include unit power, velocity gradient, unit gas flow, and digester volume turnover time. These four parameters are related and can be used to equate manufacturers' recommendations.

The unit power is defined as delivered motor W/m^3 (hp/1 000 cu ft) of digester volume. Actual energy applied, viscosity, and digester configuration are not accounted for. Several values have been suggested for unit power selection. These values range from 5.2 to 40 W/m^3 (0.2 hp to 1.5 hp/1 000 cu ft) of reactor volume. Using laboratory data, Speece (1972) predicted a level of 40 W/m^3 to be sufficient for a complete mix reactor.

The velocity gradient parameter as a measure of mixing intensity was presented by Camp and Stein (1943). It is expressed as

$$G = \left(\frac{W}{\mu}\right)^{\frac{1}{2}} \tag{4.15}$$

Where

$$\begin{aligned}
G &= \text{root-mean-square velocity gradient, s}^{-1}; \\
W &= \text{power dissipated per unit volume, Pa·s (lb-sec/sq ft); and} \\
\mu &= \text{absolute viscosity, Pa·s (lb-sec/sq ft) (for water, 720 Pa·s at} \\
&\quad \text{35°C } [1.5 \times 10^{-5} \text{ lb-sec/sq ft at 95°F]).}
\end{aligned}$$

and

$$W = \frac{E}{V} \tag{4.16}$$

Where

$$\begin{aligned}
E &= \text{power, and} \\
V &= \text{tank volume, m}^3 \text{ (cu ft).}
\end{aligned}$$

The power E can be determined from

$$E = 2.40\, P_1\, (Q) \ln \frac{P_2}{P_1} \qquad (4.17)$$

Where

Q	=	gas flow, m^3/s (cfm);
P_1	=	absolute pressure at surface of liquid, Pa (psi);
P_2	=	absolute pressure at depth of gas injection, Pa (psi).

These equations can be used to determine the necessary power and gas flow of compressors and motors for a gas injection system. Viscosity is a function of temperature, total solids concentration, and volatile solids concentration. As the temperature increases, viscosity decreases; as the solids concentration increases, viscosity increases; and as the volatile solids increase to more than 3.0%, viscosity increases. Appropriate values of G are 50 to 80 s^{-1}. The lower values can be used for a system using a single gas port or where grease, oil, and scum are suspected as potential problems.

By rearranging the preceding equations, the unit gas flow relationship to G can be solved as

$$\frac{Q}{V} = \frac{G^2 \mu}{(P_1) \ln \dfrac{P_2}{P_1}} \qquad (4.18)$$

Suggested values of Q/V for a free-lift system range from 76 to 83 $mL/m^3 \cdot s$ (4.5 to 5.0 scfm/1 000 cu ft) and 80 to 120 $mL/m^3 \cdot s$ (5 to 7 scfm/1 000 cu ft) for a draft tube system.

Turnover time is defined as the digester volume divided by the flow rate through the digester. This concept typically is used with draft tube gas and mechanically pumped recirculation systems only, where such a flow rate can actually be determined. Typical digester turnover times range from 20 to 30 minutes.

Scum, Grit, Rag, and Foam Buildup Control. Facilities must be designed to reduce or eliminate the accumulation of scum, grit, and foam in the digester tanks. These substances can interfere with digester operation by reducing the effective volume of the digesters; reducing the effectiveness of mixing, heating, and gas formation and collection; and creating operation and maintenance problems because of a failure of the digestion process.

Scum accumulation can be reduced by removing it in the clarification stages before anaerobic digestion for disposal after additional treatment. Scum removal mechanisms such as rotary screens are available with treatment or disposal of the residue at another location. Scum formation tendencies may be monitored by analyzing the influent wastewater for grease and oil concentrations. Grit should be removed at the headworks of the WWTP before the primary and secondary treatment processes. Many large-scale facilities also include primary degritters to further reduce grit loadings to the anaerobic digesters.

By maintaining a complete mixed reactor by sufficient mixing and heating, grit and scum layers can be prevented from forming in the digester. Effective mixing will keep the grit and scum dispersed throughout the tank. An excess amount of mixing may cause a foaming problem, particularly with unconfined gas mixing systems. Foam and scum buildup can be corrected by spray nozzles located in the roof structure of the digester.

Warm sprays are particularly effective in breaking up foam and scum by decreasing their viscosity and increasing their tendency to mix and disperse. Defoaming and antiscum agents that may break up the foam and scum layers are available on the market; however, those chemicals may increase the COD of the supernatant. However, maintenance of spray devices inside the enclosed vessels could be difficult.

Grit and rag removal from the digester tank can be enhanced by maintaining highly sloped floors and providing several withdrawal ports for grit removal. Access openings through the wall near the ground (where digesters extend above the ground) are helpful in grit removal when a digester is dewatered for cleaning. Tangential mixing systems may deposit grit in the tank's center.

A good example of using vessel configuration as an effective means to reduce grit and scum accumulation is the egg-shaped digester. The steeply sloped top allows for the concentration of scum in a reduced area, allowing either its removal, or its incorporation into the liquid through mixing. The steeply sloped floor concentrates grit and rags to allow their removal from the system.

Thickening. Prior thickening is beneficial to the anaerobic digestion process because it reduces biomass volume, reactor size, the volume of the supernatant, and heating requirements. Thickening methods include gravity, dissolved air flotation, gravity belt, rotary drums, and centrifugation. Because biological solids typically do not exhibit good solids–liquid gravity thickening characteristics (0.5 to 2.0% solids) in secondary clarifiers, thickening before digestion becomes necessary for economy of the reactor size. However, thickening more than 6% may cause digester mixing problems.

HEATING REQUIREMENT AND ENERGY USE. The total heating requirements for an anaerobic digester include the heat required to raise the feed to the digester operating temperature plus the summation of all the heat transfer losses through the reactor. The heat requirements for the feed is calculated as

$$Q_1 = W_f \times C_p \times (T_2 - T_1) \qquad (4.19)$$

Where

$$Q_1 \quad = \quad \text{heat required, kJ/d (Btu/d);}$$
$$W_f \quad = \quad \text{feed weight, kg/d (lb/d);}$$
$$C_p \quad = \quad \text{specific heat of water, 4.2 kJ/kg·°C (1.0 Btu/lb/°F);}$$
$$T_1 \quad = \quad \text{feed temperature during cold weather, °C (°F); and}$$
$$T_2 \quad = \quad \text{digester operating temperature, °C (°F).}$$

T_1 can be estimated from the geographical location of the digester.

Heat losses resulting from conditions at the tank roof, walls, and floor can be calculated using the following relationship:

$$Q_2 = UA \, (T_2 - T_3) \qquad (4.20)$$

Where

$$Q_2 \quad = \quad \text{heat loss kJ/h (Btu/hr);}$$
$$U \quad = \quad \text{Heat-transfer coefficient, kJ/m}^2\text{·s·°C (Btu/hr/sq ft/°F);}$$
$$T_2 \quad = \quad \text{operating temperature of digester, °C (°F);}$$
$$T_3 \quad = \quad \text{temperature outside surface of tank, cold weather average, °C (°F); and}$$
$$A \quad = \quad \text{area of exposed surface, m}^2\text{ (sq ft).}$$

The heat-transfer coefficient is a function of the type of material. Table 4.8 lists several U values for various materials and conditions.

Table 4.8   Heat-transfer coefficients for various anaerobic digestion tank materials.

Material	Heat-transfer coefficient U	
	J/m^2·s·°C	Btu/hr/sq ft/°F
Fixed steel cover, 6 mm (1/4 in.) plate	5.2	0.91
Fixed concrete cover, 280 mm (9 in.) thick	3.3	0.58
Floating cover; Downes-type with wood composition roof	1.9	0.33
Concrete wall, 370 mm (12 in.) thick exposed to air	4.9	0.86
Concrete wall, 370 mm (12 in.) thick, 25 mm (1 in.) air space and 100 mm (4 in.) brick	1.5	0.27
Concrete wall or floor, 370 mm (12 in.) thick exposed to wet earth 3 m (10 ft) thick	0.62	0.11
Concrete wall or floor, 370 mm (12 in.) thick exposed to dry earth 3 m (10 ft) thick	0.34	0.06

To calculate if sufficient energy is present to maintain the operating temperature as a result of gas production, the calculated values of the total energy required can be compared to the projected heat value of the gas assuming a combustion efficiency of 60%. The overall heat value of the gas should be greater than the heat losses or

$$G_T \times 0.60 \, Q_g > Q_1 + 24 \, Q_2 \qquad (4.21)$$

Where

G_T	=	total volume of gas produced, m^3/d (cu ft/d);
Q_g	=	heat value of gas = 24 mJ/m^3 (640 Btu/cu ft);
Q_1	=	heat required to heat feed, kJ/d (Btu/hr); and
Q_2	=	heat loss through tank, kJ/h (Btu/hr).

CHEMICAL REQUIREMENTS. Chemical feed systems may become necessary because of the changing quality and quantity of the feed. Changes in alkalinity, pH, sulfides, or heavy metal concentrations may necessitate incorporating chemical addition into the total process. The ability to feed certain chemicals such as sodium bicarbonate, ferrous chloride, ferrous sulfate, lime, and alum should be considered during the early design phase. Standby chemical metering pumps may be sufficient to satisfy operator needs. Several points of application and the associated piping necessary to accomplish feeding should be evaluated. Although pumps and other chemical feed equipment need not be installed initially, provisions to allow their future incorporation (such as pipe taps and blank flanges) should be considered during initial design.

*D*ESCRIPTION OF PHYSICAL FACILITIES

SINGLE- AND TWO-STAGE DIGESTION. Low-rate, single-stage digesters are characterized by cylindrical, square, or rectangular tanks with no auxiliary mixing or heating (see Figure 4.2). They may or may not be covered. The contents are stratified into distinct layers of scum, supernatant, active zone, and digested biosolids zone. Internal mixing is created by rising gas bubbles from within the active zone. The digested biosolids are removed from the tank bottom and feed is introduced into the active layer intermittently.

High-rate and two-stage digesters are characterized by auxiliary heating and mixing. Two-stage digestion is characterized by a second tank for solids–liquid separation and biosolids storage (see Figure 4.4). Equipment for heating and mixing varies from manufacturer to manufacturer with each having its own distinguishing features. Design and selection of a specific item should proceed only after consultation with the manufacturers and operators. High-rate two-stage systems have become the prevalent anaerobic digestion process used in recent years.

SPACE REQUIREMENTS. The selection of digester equipment and, to a certain extent, the digestion process configuration itself may be affected by physical space or land-area availability, or lack thereof. Limited space is one justification for use of single-stage, high-rate digestion with prethickening. Tank configuration and geometrics used for this process vary in their space requirements. Cylindrical tanks, especially the units that have large diameter-to-height ratios as traditionally used in the U.S., require the maximum land area for a given volume. The egg-shaped digester or one of its numerous variations, on the other hand, may be an economical choice where site area is limited or land value is high. Egg- or silo-type digesters, however, are much taller and structurally more complex to design and build.

DIGESTER COVERS. Digester designs use covered tanks to collect gas, reduce odors, stabilize internal digester temperatures, and maintain anaerobic conditions. In addition, the cover can support mixing equipment and provide access to the tank interior. Traditionally, two types of covers have been available—fixed and floating.

A fixed digester cover and its appurtenances are illustrated in Figure 4.8. These are either flat or dome-shaped and fabricated from reinforced concrete or steel. Reinforced concrete roofs are sometimes lined with polyvinyl chloride liners or steel plated to contain gas. Digester gas accumulates in the dome and is withdrawn for use or storage. Problems associated with a fixed cover may occur, especially with the introduction of air resulting in a potentially explosive gas mixture, or if positive or negative pressures are produced in the tank. This situation arises when feed and discharge rates are dissimilar, supernatant lines clog, or pressure- or vacuum-relief valves fail.

Floating-type covers are divided into two classifications: those that rest on the liquid surface and those that rest on side skirts and float on the gas (gas holders). The Wiggins-type floating cover (Figure 4.9) may have a large area above the liquid surface available for gas collection. To accomplish this, a buoyant force is developed on the outside rim of the cover that acts as a pontoon. The Downes-type floating cover (Figure 4.10) is constructed to limit the space above the liquid surface by increasing the area of cover contact with the surface. Weights are used to compensate the buoyant force to maintain the liquid level within the gas dome section. A common application for the floating-type covers is in second-stage digesters, in which volume fluctuations can be expected.

Gas holder covers are designed to provide additional gas storage space. Gas storage space allows for variation in gas production and use rates in a WWTP. Traditionally, gas holder covers have been basically a modified floating cover designed to float on gas rather than on liquid (see Figure 4.11).

Modifications include an extended rim skirt to contain the gas, and a special
guidance system to provide stability as the cover floats on gas. This type of
cover must be designed to operate under lateral wind loads and resultant over-
turning forces. A recent development in gas holder covers is the membrane
cover (see Figure 4.12). This cover consists of a support structure for a small
center gas dome and flexible air and gas membranes. An air blower system is
provided to pressurize the air space between the two membranes and vary the
air space volume. As gas storage volume increases, air volume is decreased
by bleeding air out. As gas storage volume decreases, the air space is increased by adding air from the blower. Only the gas membrane and the center
gas dome are in contact with the digester contents. The gas membrane is
made from a flexible polyester fabric similar to a lagoon liner material.

TANK CONFIGURATION AND CONSTRUCTION. Anaerobic digesters
have been built in rectangular, square, cylindrical, and egg-shaped configurations.

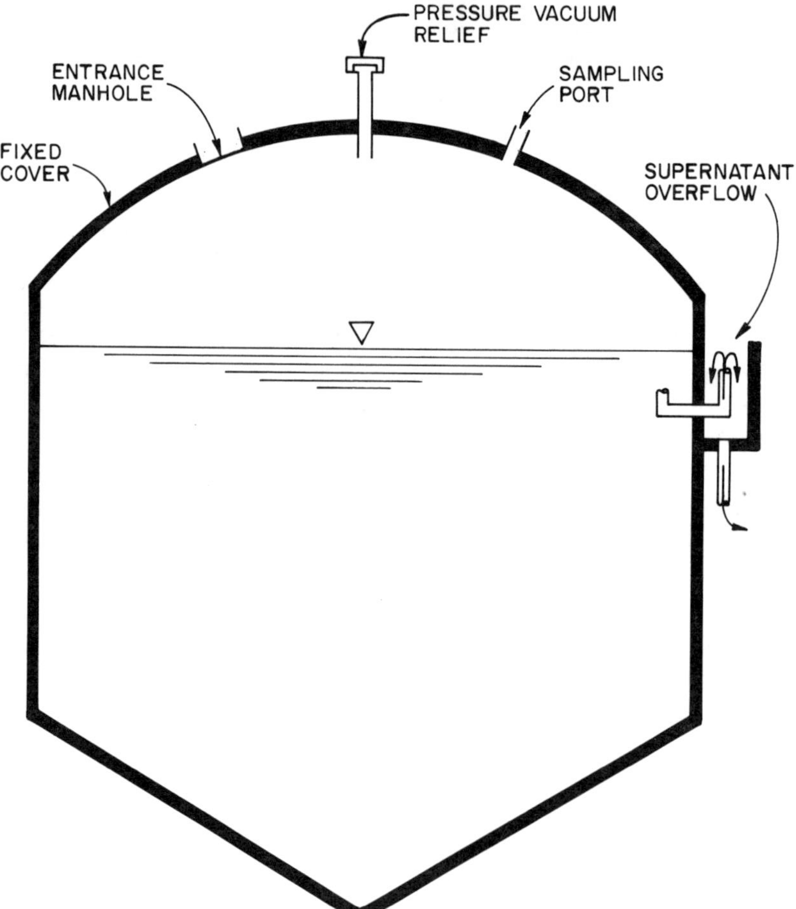

Figure 4.8   Fixed digester cover.

 Wastewater Residuals Stabilization

Rectangular tanks are used at locations where site availability is a problem. They are the least expensive to construct, but are difficult to operate because of the tendency for dead zones to develop from poor mixing characteristics.

A common configuration in the U.S. is a low vertical cylinder with a conical floor (see Figure 4.10). These circular tanks typically are constructed of reinforced concrete, with vertical side wall depths ranging from 6 to 14 m (20 to 45 ft) and diameters ranging from 8 to 40 m (25 to 125 ft) (U.S. EPA, 1979). Conical-shaped bottoms are preferred for cleaning purposes, with slopes varying between 1:3 and 1:6 (WPCF, 1987). Slopes greater than 1:3, although desirable for grit removal, are difficult to construct and also difficult to work on when cleaning. However, milder slopes typically are too flat to provide any significant improvement over flat bottom designs. The floors can either contain one central withdrawal pipe or may be divided into pie-like sections each equipped with a separate withdrawal pipe (waffle bottom digester). The latter are more costly to construct than traditional conical designs, but may reduce cleaning costs and frequency. Where necessary, cylindrical digesters have been insulated using brick veneer and an air space, earth fill, polystyrene plastic, fiber glass, or insulation board.

A tank configuration that is relatively new in the U.S., but has been used extensively in Europe, is the egg-shaped digester (Figure 4.13). The steeply sloped

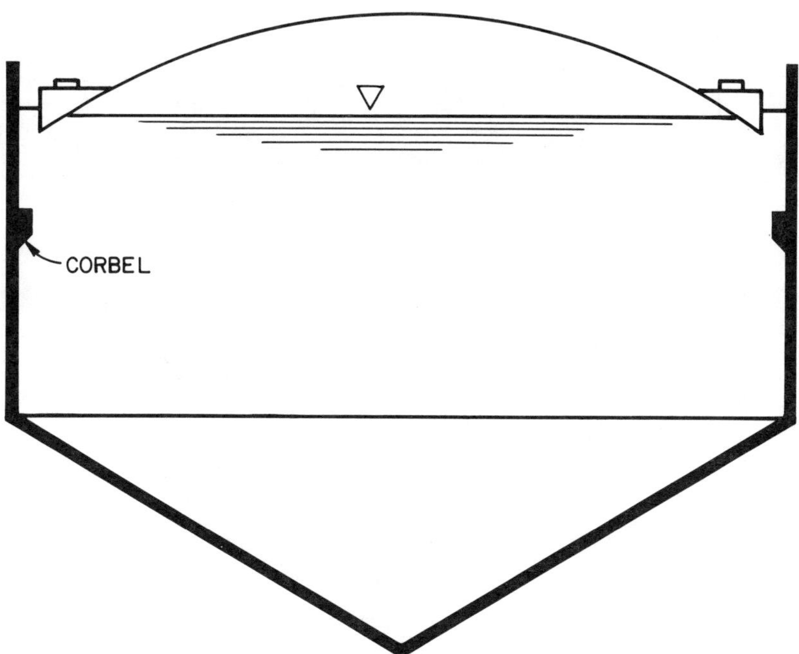

Figure 4.9   Floating digester cover.

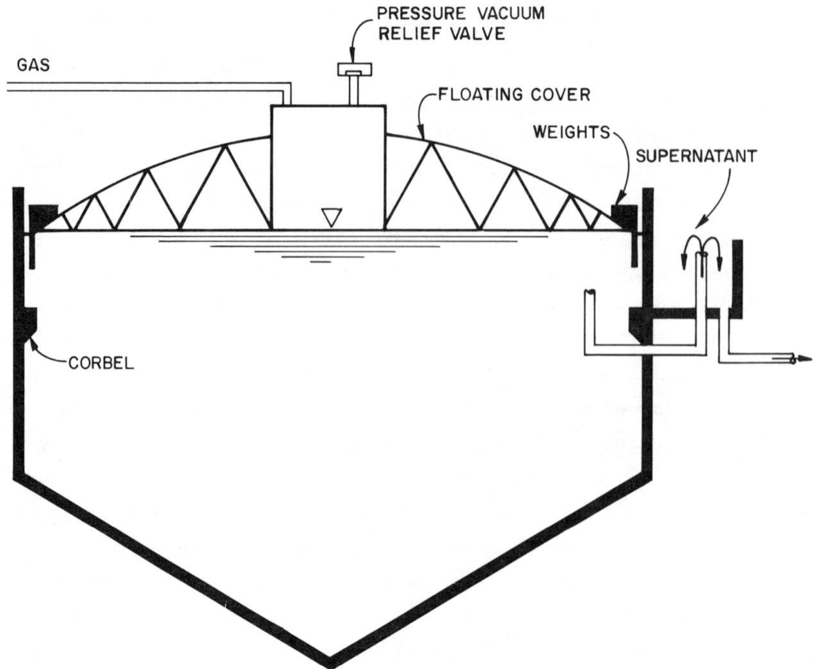

Figure 4.10   Cylindrical tank with Downes-type floating cover.

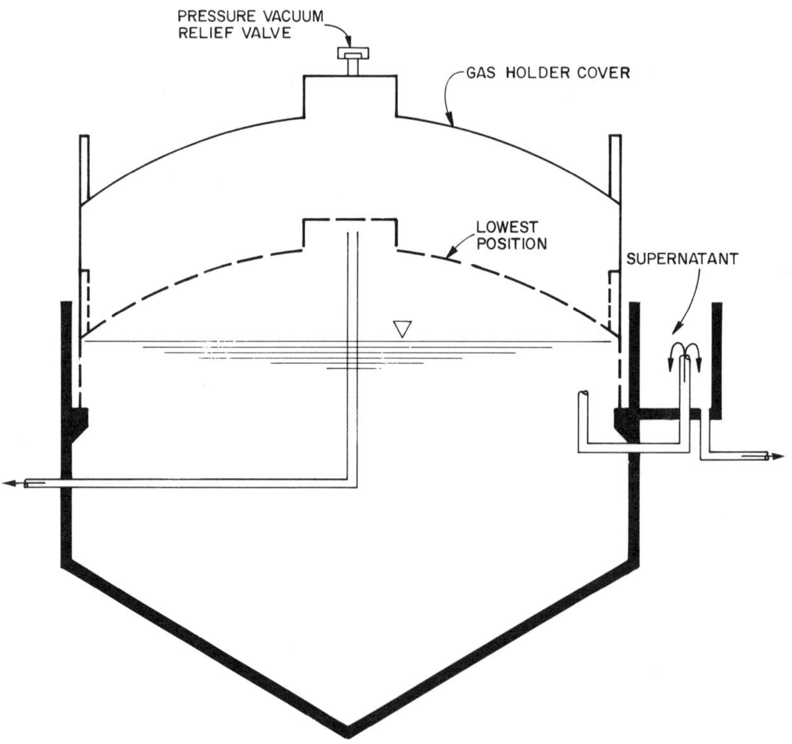

Figure 4.11   Gas holder floating cover.

top and bottom cones help in eliminating scum and grit buildup problems, which in turn eliminates or reduces the necessity of digester cleaning. Mixing requirements of an egg-shaped digester are considerably lower than for a conventional shallow cylindrical vessel. In the latter, most of the mixing energy applied goes to maintaining grit in suspension and to control scum buildup.

There are three basic types of mixing systems used with egg-shaped digesters: unconfined gas mixing, mechanical mixing using an impeller and a draft tube arrangement, and external pumped recirculation (see Figure 4.14) (Stukenberg *et al.*, 1990). Most egg-shaped digesters are also fitted with a gas lance or hydraulic jet near the bottom of the lower cone section to facilitate occasional stirring of grit that may have accumulated. Any or all of the mixing systems may be provided in any one digester, and all may be operated during any one day, although gas mixing and mechanical draft-tube mixing are seldom practiced at the same time. Egg-shaped digesters may be constructed of steel or concrete. The outside surfaces are insulated and covered, typically with aluminum cladding for protection and aesthetics. The egg-shaped digesters are higher in capital costs compared with cylindrical or rectangular units because of the specialized construction techniques required. However, these higher capital costs can be offset by the reduction of land requirements because of their tall and narrow configuration, and by the elimination or reduction of cleaning requirements.

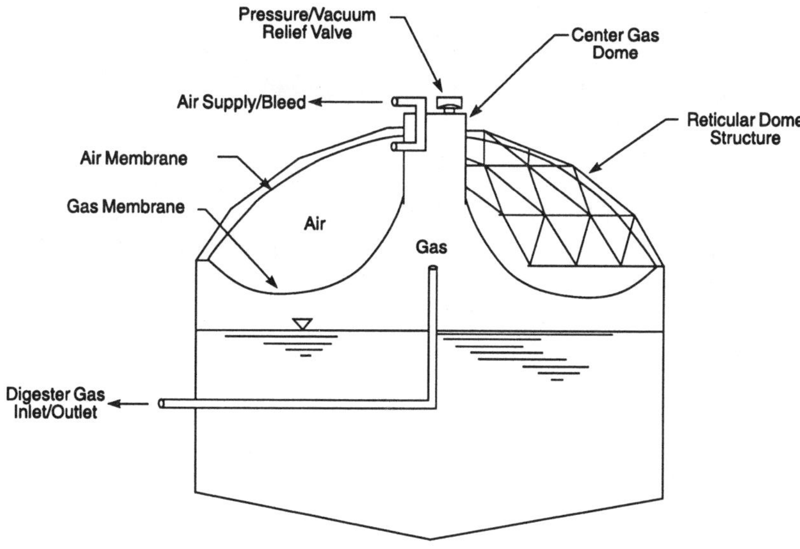

Figure 4.12   Membrane gas holder cover.

PUMPS AND PIPING CONFIGURATIONS. Pumps typically used for transporting the feed to and biosolids from the digester are piston, progressing cavity, vortex centrifugals, and nonclog centrifugals. These pumps are capable of handling grit and most solids. The pump size is sometimes limited by the suction pipe diameter. Typical solids concentrations range from 2 to 6% for combined primary and secondary and 4 to 10% for primary. A nonclog centrifugal pump has an open impeller to accommodate solids, greater tolerance between impeller and casing, and wear-resistant bearings, shafts, and seals. A vortex or torque centrifugal pump is characterized by an open-faced suction impeller recessed from the suction intake. This condition results in the majority of the fluid passing through the pump without contacting the impeller and causing abrasive wear. However, this geometry results in a less efficient pump.

One important factor in selecting a pump for transporting solids is the construction material used inside and outside the pump. Inside materials should

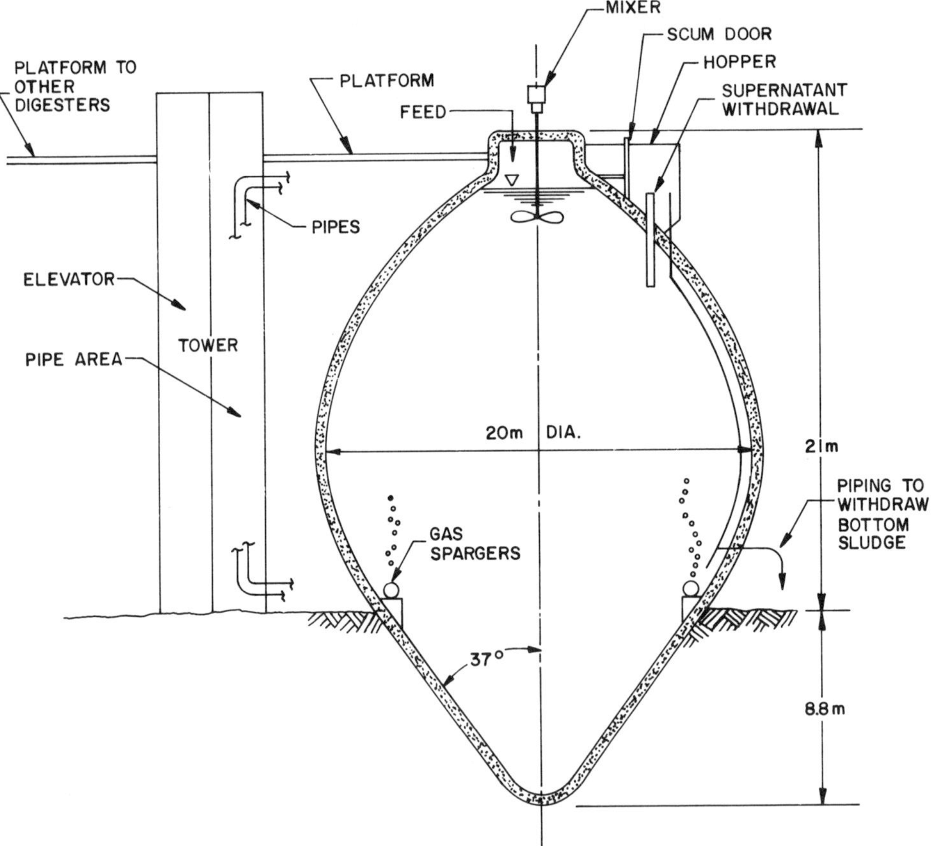

Figure 4.13   Egg-shaped anaerobic digester.

be resistant to abrasive action, corrosion, and cavitation. Nickel and chrome impellers and casings are typically used for their resistance characteristics. Polytetrafluoroethylene and other plastics can be used for the stator material in progressing cavity pumps and the rotor can be tool steel. Pump exteriors should be maintained by painting to prevent corrosion. Another important factor in selecting a pump is how easily it can be serviced and dismantled to allow the removal of rags and other debris from the pump.

Piping configurations should be designed to promote maximum flexibility for feeding, recirculating, and discharging solids. Piping should be arranged to provide several points for solids feed, solids withdrawal, and supernatant withdrawal. Because solids pumping is characterized by low velocities and possible solids accumulation in pipelines, provisions should be made for the cleanout and backflushing of lines (using treated effluent wastewater when available). Selection of valves and valve placements must also be considered. Easy access to valves is desirable with provisions for manual operation. Provisions should be included to allow the isolation of all tanks and pumps for maintenance and safety purposes.

Piping should be arranged to accommodate the following operating modes (for two-stage digestion): transferring the biomass by gravity from the first stage to the second, pumping the biomass from one digester to another, multiple ports for supernatant withdrawal, recirculation systems using several suction/discharge ports, and backup pumps with appropriate piping.

GAS-HANDLING SYSTEM. The necessary pieces of equipment for a typical gas-handling system for an anaerobic digester include those for collection, transportation, metering, storage, and safety (Figure 4.15).

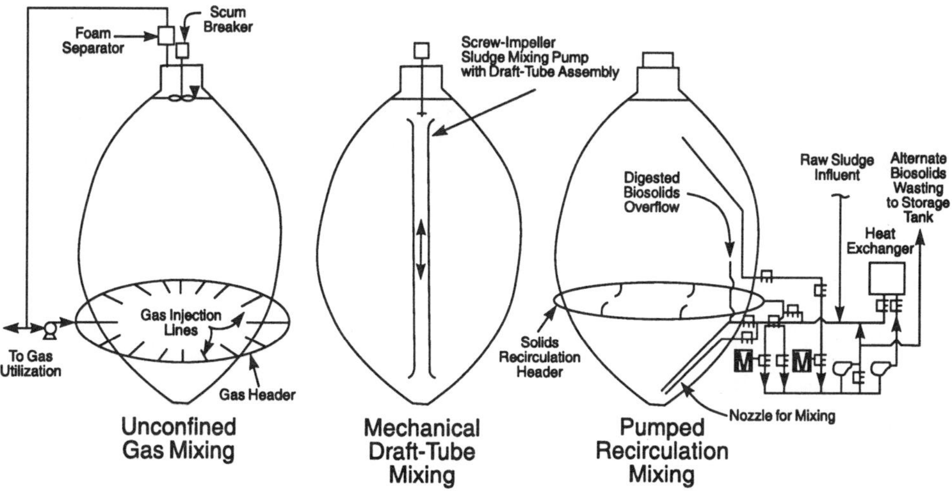

Figure 4.14   Mixing systems for egg-shaped anaerobic digesters.

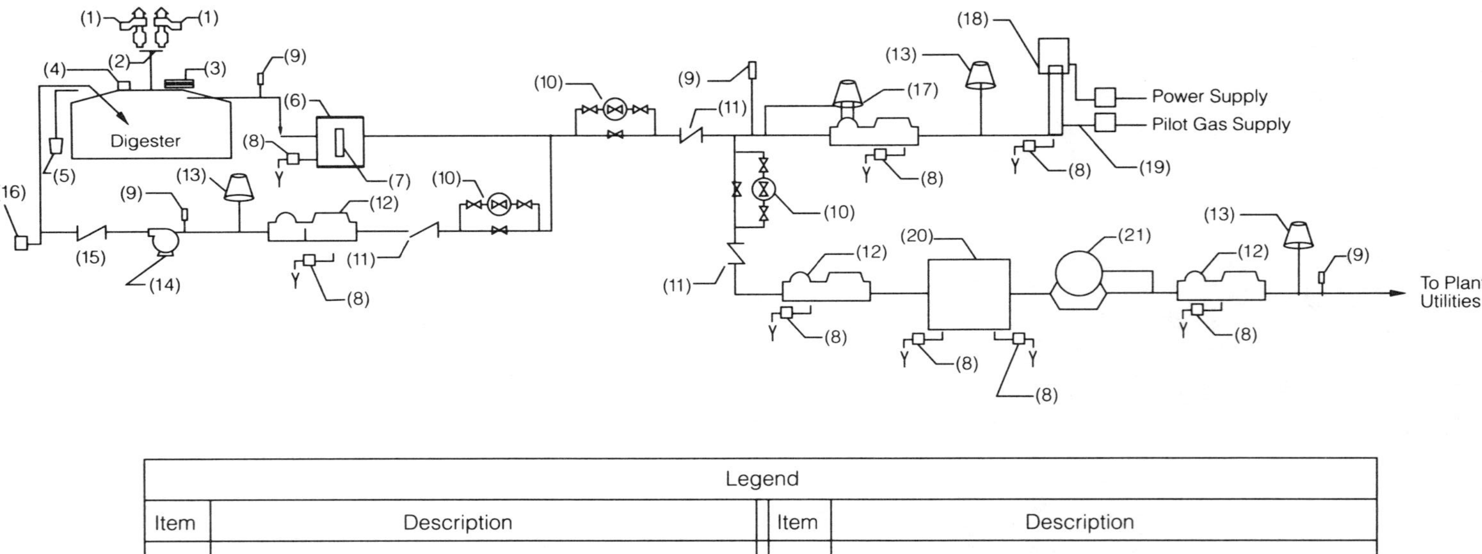

Figure 4.15　Schematic diagram of a gas control system (single-digester, gas-handling system) (WEF, 1992).

The pressure- and vacuum-relief valves are necessary to alleviate negative and excessive pressures developed under the digester cover during gas formation. At set pressures, gas is allowed to escape or enter depending on conditions within the tank. Excess solids withdrawal can cause vacuum conditions, while plugged solids or gas withdrawal lines can cause high-pressure conditions. Condensed water and sediment are removed from the gas stream by sediment tank and drip-trap assemblies located throughout the system at low points. Drain traps should be checked daily. The use of automatic drains from the traps has been discouraged because of the potential for gas leaks.

A flame-trap assembly is installed between any potential flame source and the digester to prevent the flame from reaching the digester tank. Because of impurities, these traps should be examined monthly. Pressure regulators should be used to adjust gas pressures to those needed for gas utilization systems such as boilers. Gas meters should be used to measure the quantity of gas produced, wasted, and used. Meter operation should be checked daily.

Manometers installed in the gas piping should be used to monitor operating pressures. Compressors should be used to pump gas under pressure for storage up to 350 kPa (50 psig). High-pressure gas storage spheres or low-pressure gas holder floating covers have been used to store gas for variations in gas use. Excess gas can be wasted by use of a waste gas burner. The waste burner can be ignited automatically by electric pilot ignition or pilot light when the storage vessel is filled to capacity.

Digester gas can be used as fuel for internal combustion or gas turbines that are, in turn, used for pumping, operating blowers and compressors, and generating electricity. Engine coolant and exhaust gases may be also used as a heat source. The production of both thermal and mechanical energy from digester gas is often referred to as cogeneration. However, depending on the gas quality, removing hydrogen sulfide could be required to protect engines and other gas-handling components from corrosion. Applications with hydrogen sulfide concentrations in excess of 100 ppm in the gas should consider hydrogen sulfide removal equipment.

MIXING EQUIPMENT. Digester mixing systems can be classified into the following four categories:

- Confined gas injection systems,
- Unconfined gas injection systems,
- Mechanical stirring systems, and
- Pumped mixing systems.

Confined gas injection systems using draft tubes are a common type of digester mixing system designed today. They can provide enough mixing to ensure a completely mixed process. A draft tube gas recirculation system consists of a series of large-diameter tubes into which digester gas is released, causing the biomass to rise and mix as it approaches the liquid surface of the tank. The number of draft tubes in a digester will depend on the digester dimensions. Typically, more than one draft tube is provided for digesters with diameters greater than 18 m (60 ft) (Liptak, 1974). Compressed gas is released inside the draft tube by means of top entering lances, or laterally through the draft tube wall near the bottom of the unit. Figure 4.16 shows a single draft tube installation mounted from the roof. However, they can also be mounted on the digester floor with supports. The system is supported by the required compressors and controls. The three types of compressors typically used are the rotary lobe, rotary vane, and liquid ring. The draft tubes typically are constructed of steel plate with typical diameters ranging between 0.5 and 1.0 m (20 and 40 in.). A heating jacket can be installed around the draft tube to provide mixing and heating (see Figure 4.16).

Mechanical stirring systems use rotating impellers to mix the digester contents. The mixers may be either low-speed turbines or higher-speed propeller mixers installed in draft tubes. The draft tubes may be either internally or externally mounted on the digester. The flow pattern for mechanically stirred and pumped mixing systems is from top to bottom in contrast to gas systems, which mix from bottom to top. Disadvantages of mechanical stirring systems include sensitivity to liquid level and clogging of impellers with rags (U.S. EPA, 1979).

In pumped mixing systems, externally mounted mixing pumps withdraw biomass from the top center of the tank and reinject it through nozzles mounted tangentially at the bottom of the tank. Scum breaker nozzle(s) are also provided at the liquid surface for intermittent use in breaking up scum accumulations. A high-flow, low-head, solids-handling pump, such as an axial

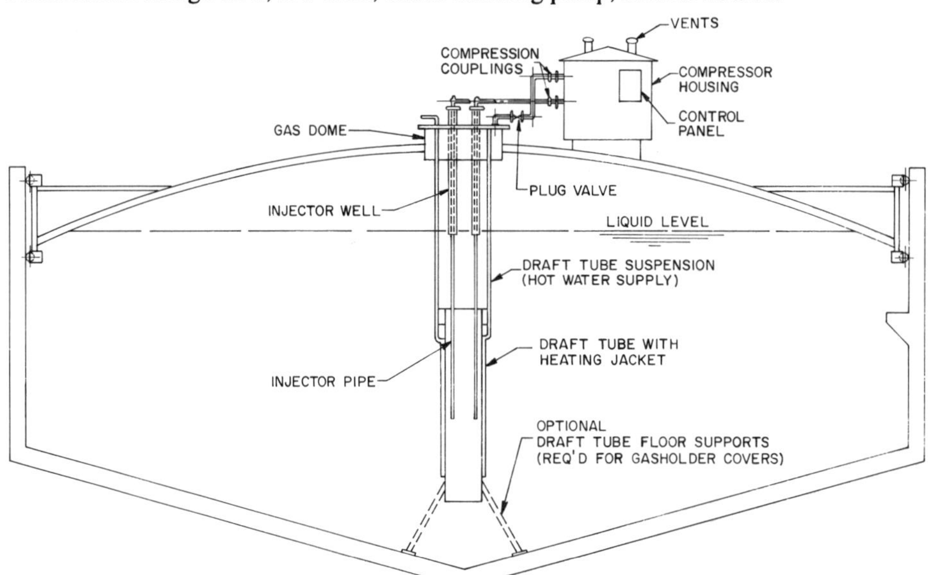

Figure 4.16 Single draft tube mixer with heating jacket.

flow, mixed flow, or screw centrifugal pump, is used. Pumps often are belt-driven to permit speed changes if digester solids contents vary.

Multiple-drop lance gas recirculation is a typically used, unconfined system consisting of several gas injector pipes located throughout the tank. Gas may be discharged continuously through all pipes, or sequentially from one injector pipe through another by means of a rotary valve. The rotary valve operation typically is controlled automatically by use of preset timers. The lances or gas injection pipes are located approximately two-thirds out from the digester center. To ensure mixing near the center, an additional lance may be located several meters from the center. Again, the system requires compressors and control equipment. Figure 4.17 details a profile view of a multiple sequential discharge lance system. The gas discharge lines are at least 50 mm (2 in.) in diameter and provisions must be made for condensate accumulation. The submerged distance of the lances is a major factor in determining gas flow rates. Figure 4.18 shows a plan view of the dropped lance system using six lances for a tank diameter of 13 to 15 m (41 to 50 ft).

Another unconfined gas injection system is based on the use of floor-mounted diffuser boxes, arranged in a circular pattern and mounted to short concrete pedestals (see Figure 4.19). The number of diffuser boxes depends on the size and volume of the digester tank. Each diffuser is supplied with compressed gas by a separate gas line from a compressor. In contrast to other gas injection systems, the mixing geometry is independent of floating cover height. These systems may be difficult to maintain because they are attached permanently to the digester bottom.

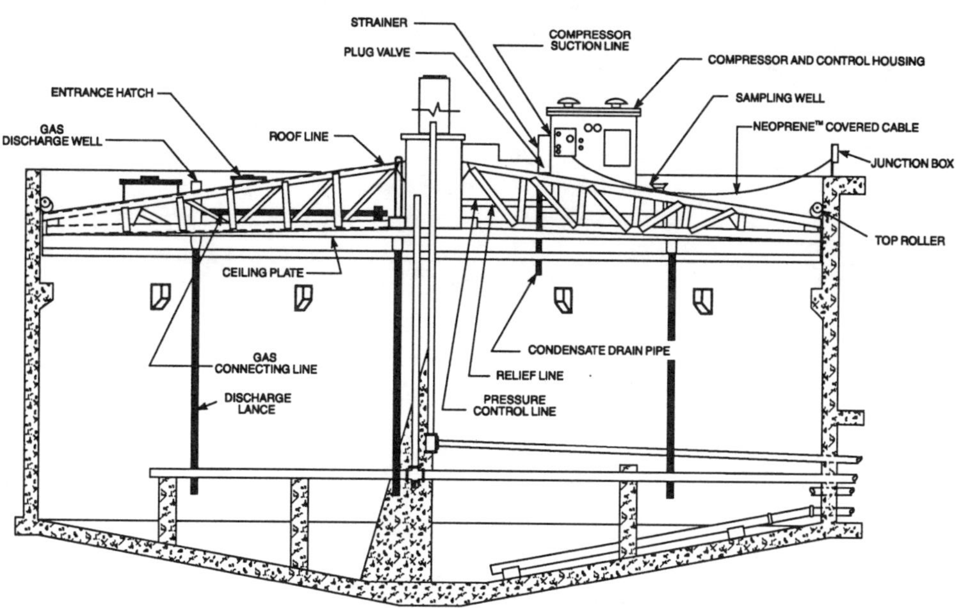

Figure 4.17   Multiple sequential discharge lance mixer.

HEATING EQUIPMENT. Both internal and external heating equipment have been used to maintain operating temperatures. Older digesters may contain internal heating coils attached to the walls in which hot water circulates. These coils are subject to fouling, causing a decrease in the heat transfer. Maintaining these coils requires the operator to remove the digester from operation and empty its contents. A heat jacketed draft tube mixer may also be used to internally heat the sludge (Figure 4.16). However, internal heating systems are seldom used given the difficulty in providing maintenance within the vessel.

Water bath, tube-in-tube, and spiral plate external heat exchangers can be used for anaerobic digesters. Operation of a water bath external heat exchanger involves the circulation of the biomass by pumping through a heated water bath. The heat-transfer rate is increased by pumping hot water in and out of the bath (see Figure 4.20). The use of recirculation pumps for external heat exchangers allows the feed to be heated before introduction to the digester, although this is not essential to adequate heating.

The tube-in-tube and spiral plate heat exchangers are similar in design. A tube-in-tube exchanger consists of two concentric pipes (one containing the biomass, the other containing hot water) flowing countercurrently. Spiral plate heat exchangers are composed of two long strips of plate wrapped to

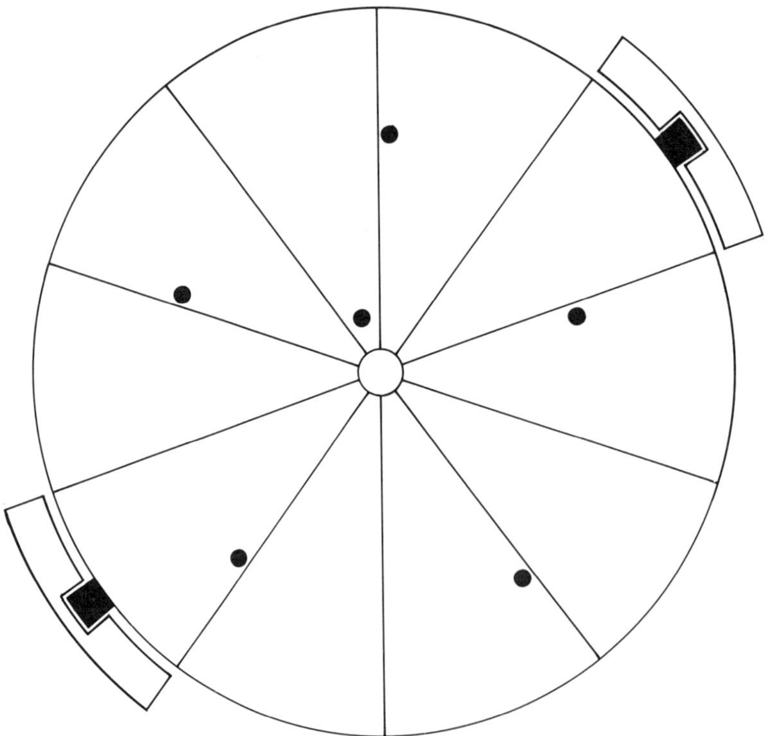

Figure 4.18   Relative position of sequentially operated drop lances.

form a pair of concentric spiral passages (Figure 4.21). The flow scheme for a spiral-heat exchanger is also countercurrent. The wrapped plates are designed with maximum clearances to resist blockage and assist maintenance. Water temperatures are kept below 68°C (154°F) to prevent caking.

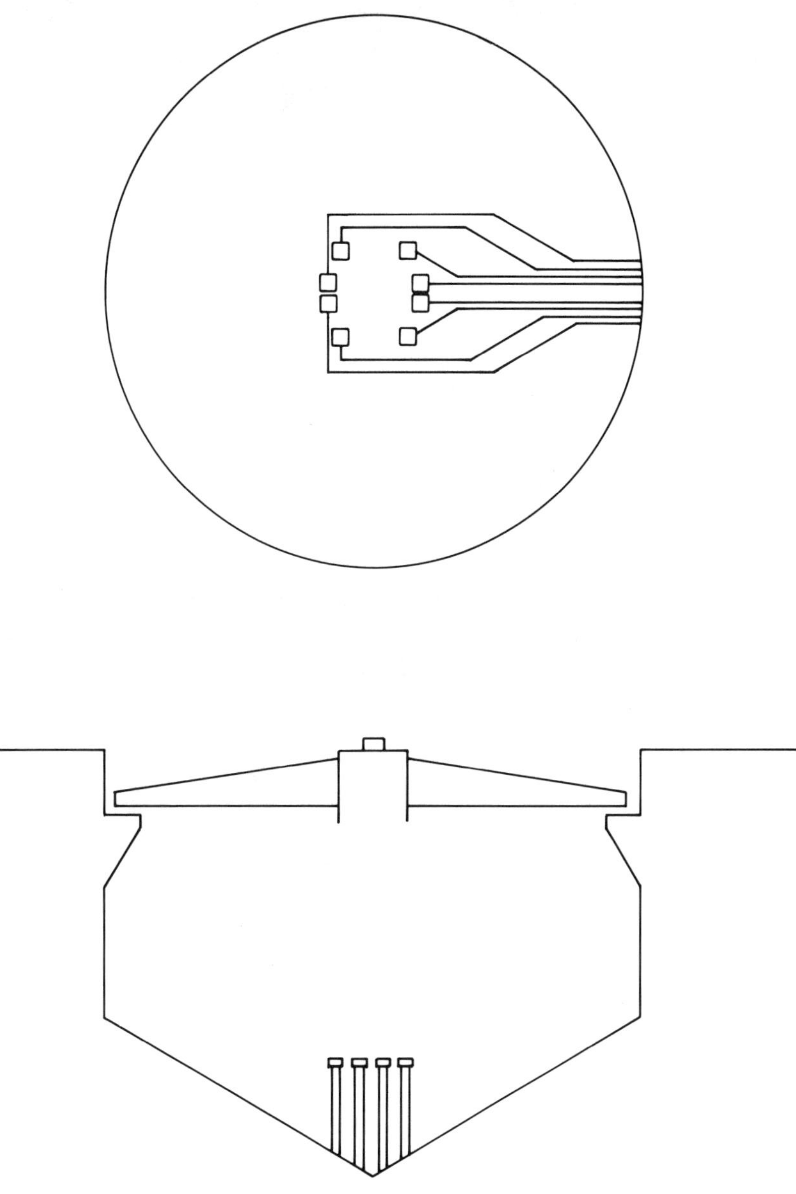

Figure 4.19   Floor-mounted gas diffuser boxes.

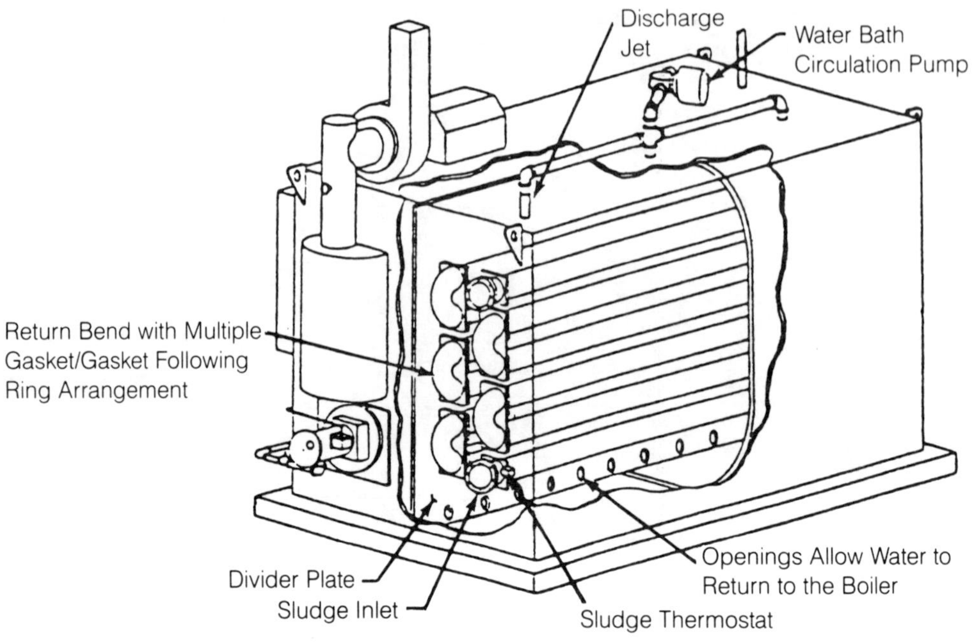

Figure 4.20   Cutaway view of a water bath heat exchanger (WPCF, 1987).

Heat-transfer coefficients for external heat exchangers range from 0.9 to 1.6 kJ/m^2·s·°C. Transfer coefficients for internal coils used for heating range from 85 to 450 kJ/m^2·s·°C (15 to 80 Btu/hr/sq ft/°F) depending on the solids content of the biomass.

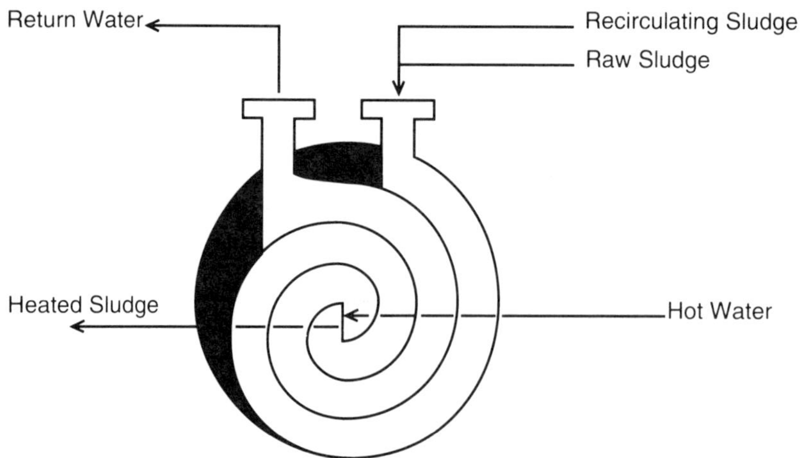

Figure 4.21   Schematic of spiral heat exchanger (WEF, 1992).

Boilers typically are used to supply heat to the circulating water in the heat exchangers. These boilers can be fueled exclusively by digester gas, but provisions can also be included for use of an auxiliary fuel such as natural gas or fuel oil.

CHEMICAL FEED SYSTEMS. Ideally, digester chemical feed systems should be integrated to the overall chemical feeding and metering equipment for an entire WWTP. This would permit optimal use of installed equipment, because addition of chemicals to a digestion process may not be routinely necessary.

Chemical addition may become necessary for two main reasons: pH/alkalinity control and inhibition/toxicity control. Sodium bicarbonate, sodium carbonate, and lime are examples of chemicals used to provide alkalinity. Ferrous chloride, ferrous sulfate, and alum may be used for precipitation/coagulation of inhibitory substances or to control offgas hydrogen sulfide content.

Chemical feeding facilities include supply vehicle access, unloading and storage equipment, transfer and dissolution/dilution equipment, and metering/delivery equipment. Depending on the chemical supplier, unloading facilities might include provisions of weighing scales, hoses, hoppers, chutes, compressed air/vacuum, and unloading pumps. Most suppliers can provide detailed information on materials, handling, safety, and storage (temperature, ventilation, recirculation, emergency equipment) requirements for each chemical at various concentrations.

Dissolution/dilution to the proper application concentration can be accomplished with flow-proportioning instrumentation. Such instrumentation maintains an operator-specified ratio between flows of a stock chemical and makeup water.

Chemical metering is most easily accomplished by using metering pumps to pump undiluted stock chemicals at controlled rates.

Depending on the kinetics of the desired precipitation reaction and the efficiency of digester equipment, additional mixing may also be required for effective use of chemicals.

CLEANING AND SAFETY. The decision to clean a digester tank is based on several factors including the accumulation of grit and scum and the reduction of the effective volume of the digester, the condition of internal heating and mixing equipment, the availability of alternative solids handling equipment, and tank structure. Digester tanks and mixing and heating equipment should be designed for easy access during cleanout operations. Access manholes, 0.9 m (36 in.) in diameter, should be provided on the top and sides of the digester tank. These manholes should be large enough to enable an operator to use grit and scum removal equipment. Heating and mixing equipment will have to be maintained throughout the life of the digester. Interior units that are removable will be more easily maintained. However, exterior equipment typically is preferred. Digester cleaning can be done using in-house staff or a contractor specializing in these services.

In conjunction with cleaning the digester, safety is of prime importance. Anaerobic digesters are confined spaces and all systems must be designed to ensure maximum safety for the operations and maintenance staff. This includes specifying the appropriate personal safety equipment necessary for entering a digestion tank for inspection and cleanout. Vacuum- and pressure-relief valves, flame traps, and automatic thermal shutoff valves must be designed into the gas collection and piping system. The proper enclosures and ventilation must be provided. Additional safety features are provided by the National Electric Code (NFPA, 1993), Recommended Standards for Sewage Works (Health Educ. Serv. Inc., 1978), and *Safety and Health in Wastewater Systems* (WEF, 1994). Several pieces of equipment are available to perform a safe cleaning operation. Table 4.9 lists the equipment available for cleaning and safety. Depending on the size of the operation, not all items contained in the list are necessary. Before entering the digester, the air composition inside must be monitored for oxygen deficiency and life-threatening gases. The

Table 4.9   Digester cleaning and safety equipment.

Sludge line valves

Sludge line (permanent)

Sludge line (temporary)

Digester access

Explosionproof vent fan

Explosive level meter

Safe ladder

Self-contained breathing apparatus

Safety harness

Nonskid boots

Explosionproof lights

Water source

Washdown hose

Nozzle with shutoff

Wash water pump

Fixed sludge pump

Portable sludge pump

Turret nozzle

Tripod or hoist

Tank truck

Crane

safety equipment is installed to prevent falls, infection, injuries during equipment operation, suffocation and asphyxiation, and explosions.

*O*PERATIONAL CONSIDERATIONS

PROCESS MONITORING. The anaerobic digestion process is sensitive to changes in operating conditions. These changing conditions, if not controlled, may result in digester upsets and failure. Proper monitoring techniques will promote successful operation and ensure process stability and methane production. All process streams should be available for sampling and analysis. Feed, digested biosolids, supernatant, digester gas, and the heating fluid (hot water) should be analyzed for various constituents and physical conditions. Sampling ports should be incorporated into the design to ensure adequate access for sampling. Feed should be analyzed for total solids, volatile solids, pH, alkalinity, and temperature; digester content and effluent biosolids for the above plus volatile acids; digester gas for volume and percent methane, carbon dioxide, and hydrogen sulfide composition; supernatant for pH, BOD, COD, total solids, total and ammonia nitrogen, and phosphorus; and heating fluid for total dissolved solids and pH.

All streams should be monitored for volume of flow per day by accurate meters. For digesters that produce supernatant, it is important to quantify this stream. This information should be used to compute total solids quantity in the system and subsequently the SRT. Additional monitoring requirements such as those for toxics should be decided on a case-by-case basis.

PROCESS INHIBITION. U.S. EPA studies that investigated the fate of priority pollutants have confirmed their presence in wastewater residuals. Many of these may be inhibitory to the anaerobic digestion process to varying degrees depending on the range of concentration. Additional substances, such as sulfides, ammonia, volatile acids, and light metal cations, can also inhibit process performance. The degree of this toxicity depends on several criteria. Only the soluble portion of a substance is available to the organisms for incorporation; however, influent solids containing toxic substances may be solubilized. If dosed at small enough increments, many microorganisms can adapt themselves to relatively high concentrations of toxic substances, and the presence of additional substances may negate or increase the toxicity of a particular compound. However, the presence of toxic substances has been reduced as the result of the implementation of industrial pretreatment programs nationwide. Also, beneficial use requirements for biosolids have placed limits on many compounds also thought to be potential inhibitors of the digestion process.

As previously stated, volatile acid concentrations may become inhibitory if the pH is lowered. Sufficient levels of alkalinity and an acceptable volatile acids to alkalinity ratio can ensure good operation. Addition of bicarbonate is useful in controlling upsets as long as certain light metal cation concentrations associated with bicarbonate salts are not exceeded (for example, calcium, sodium, and potassium).

The presence of light metal cations can be stimulatory, moderately inhibitory, or acutely inhibitory (see Table 4.10). Calcium, magnesium, potassium, and sodium concentrations should be kept in the stimulatory range. The harmful effects of some cations may be counteracted or multiplied by the presence of others. Kugelman and McCarty (1965) list synergistic and antagonistic effects of different cations (see Table 4.11).

Toxic effects from soluble sulfide have been reported at concentrations in excess of 200 mg/L (Lawrence and McCarty, 1965). The sulfide concentration in the digester is a function of influent sulfur concentration, pH, gas production rate, and heavy metal concentration. Soluble sulfides may be scavenged by heavy metal precipitation or iron salt addition.

Heavy metal concentrations pose a significant threat to the successful operation of solids handling and disposal facilities. They can be inhibitory to process function and also limit the mode of disposal. The amount of heavy metals present in the feed depends on the treatment process, the metal concentration of raw wastewater, and other factors such as pH. The mechanism by which heavy metals are incorporated can be by sorption or precipitation. At pH values of more than 7.2, precipitation may be the mechanism. Heavy metal ions will adsorb to sludge particles, sometimes making it difficult to distinguish soluble versus insoluble levels. Because of this measurement problem, exact levels of metal toxicity are difficult to compute and are highly pH dependent.

Table 4.12 lists concentrations at which heavy metals have been found to be inhibitory to anaerobic digestion. Data concerning threshold inhibitory concentrations are given in Table 4.13.

Toxic effects from ammonia can be caused by both of its chemical forms, either as dissolved ammonia gas or ammonium ion. The concentration of each form is pH-dependent with ammonia gas predominant at levels greater

Table 4.10 Stimulatory and inhibitory concentrations of light metal cations (McCarty, 1964).

	Concentration, mg/L		
Cation	**Stimulatory**	**Moderately inhibitory**	**Strongly inhibitory**
Calcium	100–200	2 500–4 500	8 000
Magnesium	75–150	1 000–1 500	3 000
Potassium	200–400	2 500–4 500	12 000
Sodium	100–200	3 500–5 500	8 000

Table 4.11   Synergistic and antagonistic cation combinations (Kugelman and McCarty, 1965).

Toxic cations	Synergistic cations	Antagonistic cations
Ammonium	Calcium, magnesium, potassium	Sodium
Calcium	Ammonium, magnesium	Potassium, sodium
Magnesium	Ammonium, calcium	Potassium, sodium
Potassium		Ammonium, calcium, magnesium, sodium
Sodium	Ammonium, calcium, magnesium	Potassium

than pH 7.6. Inhibitory concentrations of total ammonia nitrogen are 1 200 mg/L and higher at a pH of more than 7.4. However, pilot-plant and laboratory studies indicate that significantly higher ammonia nitrogen concentrations can be tolerated if the methane-forming bacteria are properly acclimated (Kroker *et al.*, 1979). Combined feeds with a content greater than 30% and a solids concentration more than 6% may contain high ammonium concentrations that could inhibit the anaerobic digestion process.

ODOR CONTROL. Hydrogen sulfide gas and gaseous ammonia are the main odor-causing byproducts of anaerobic digestion. Also, hydrogen sulfide production reduces the methane content of digester gas, which results in a lower heating value during combustion.

Hydrogen sulfide production cannot be eliminated completely, but would not cause an odor problem in a leak-free, gas-handling system. Floating covers are an exception because they allow a liquid interface with the atmosphere.

Table 4.12   Total concentration of individual metals required to severely inhibit anaerobic digestion (DeWalle *et al.*, 1979, and Mosey, 1976).

Metal	Concentration in digester contents		
	Metal as % of dry solids	mM metal/kg dry solids	Soluble metal, mg/L
Copper	0.93	150	0.5
Cadmium	1.08	100	—
Zinc	0.97	150	1.0
Iron	9.56	1 710	—
Chromium			
+6	2.20	420	3.0
+3	2.60	500	—
Nickel	—	—	2.0

Odor problems can be associated mainly with predigestion thickening, digester supernatant (if supernatant is recycled), postdigestion dewatering, and intermediate and final storage. High-strength supernatant streams can cause odors in other WWTP processes as well. Hydrogen sulfide can also cause corrosion problems when the digester gas is used in boilers or engine generators.

Use of dissolved air flotation thickening may reduce predigestion odors compared with gravity or gravity belt thickening or centrifugation. Predigestion thickening can also help eliminate the need for supernating and the associated odors. Covered storage vessels with pressure-relief discharges connected to the digester gas-handling system can eliminate storage odor problems.

Odors from postdigestion dewatering operations are the most difficult to control/contain. Options include addition of odor control chemicals and installation of odor scrubbing equipment to treat the ventilation air around the process. Chemical addition limits final disposal/reuse options, while scrubbing equipment has higher capital and operating costs. The choice largely depends

Table 4.13   Toxic pollutants inhibitory to anaerobic digestion (Anthony and Breimhurst, 1981).

Toxic pollutants	Threshold inhibitory concentrations,* mg/L
Acrylonitrile	5 (S)
Arsenic	1.6 (S)
Benzidine	5 (S)
Cadmium	0.02 (S)
	< 20 (T)
Carbon tetrachloride	10 – 20 (S)
Chloroform	10 – 16 (S)
Chromium	
Chromium (+6)	5 – 50 (S)
	110 (T)
Chromium (+3)	50 – 500 (S)
	130 (T)
Copper	1 – 10 (S) 40 (T)
Cyanide	4 (S) 1 – 2 (S)
Hexachlorocyclohexane	48 (S)
Lead	340 (T)
Mercury	13 – 65 (S)
Nickel	10 (T)
Pentachlorophenol	0.4 (S)
Trichloroethylene	20 (S)
Zinc	5 – 20 (S)
	400 (T)

* (T) = total and (S) = soluble.

on location, the community, and its financial resources. For dewatering options, centrifuges offer the best advantage for odor control, given that their enclosed and compact design facilitates offgas collection for treatment.

SIDESTREAMS. After the decision has been made to incorporate anaerobic digestion to the solids handling train, the quality and treatment of the supernatant sidestream must be considered. In recent years the practice of withdrawing and recycling supernatant to the head of the WWTP has been discontinued, particularly in systems digesting activated sludge that exhibit poor solids–liquid separation. Even WWTPs without activated sludge may not benefit from supernatant recycle, if significant volume reduction is not achieved in the digested biosolids.

Few digester facilities practicing supernatant recycle attain a digested biosolids concentration equal to, or higher than, the digester feed solids concentration. If the digested biosolids do not attain close to their original concentration after supernatant decanting, then solids–liquid separation may be too poor to justify continued supernatant withdrawal. Supernatant quality is illustrated in Table 4.14 for primary, trickling filter, and activated-sludge WWTPs. The values for BOD, COD, total solids, total phosphorus, organic and ammonia nitrogen, and suspended solids are greater than those values for domestic wastewater. As a result, the total daily load of a particular parameter may be substantially affected by this supernatant flow. It should be recognized, however, that even if digester decanting is not practiced, other high-strength streams such as those resulting from digested biosolids dewatering will have similar characteristics.

Where supernatant recycle is determined as beneficial, several techniques have been developed to counteract any problems that may occur because of highly concentrated supernatant flows. These include modifying the upstream solids treatment train, providing a separate supernatant treatment facility, and designing (or modifying) the wastewater treatment train to accommodate the supernatant.

Possibly the most beneficial upstream treatment modification is thickening before digestion. This may reduce the quantity of the supernatant and may even eliminate the need for recycling supernatant. Thickening is also beneficial because its net effect is to reduce the necessary digester volume and heating requirements. Thickening may be accomplished by either gravity, dissolved air flotation, gravity belt, or centrifugation (WEF, 1992, and WPCF, 1980).

Facilities for solids, nitrogen, and phosphorus removal may have to be expanded if treatment of the sidestream is to be accomplished in the wastewater treatment train. The sidestream typically is returned upstream of the primary clarifier; therefore, the primary clarifier, the biological treatment facilities, and the secondary clarifier will have to be designed to handle the additional organic load.

Several systems such as coagulation, aeration, biological filters, storage in lagoons, and chlorine oxidation have been used to treat the supernatant before its introduction to the primary clarifier. Varying degrees of success have been reported for these methods. Lime and ferric chloride were used successfully by one researcher (Rudolphs and Fontenelli, 1945). Before choosing any of

Table 4.14 Supernatant characteristics of high-rate, two-stage mesophilic anaerobic digestion at various plants (U.S. EPA, 1979).

| | Concentration,[a] mg/L | | | | | | | | | |
| Parameter reference | Primary sludge | | Primary and trickling filter sludge | | | Primary and activated sludge | | | | |
	(15)	(16)[b]	(17)	(15)	(16)[c]	(17)	(17)	(15)	(14)	(16)[b]
Total solids	9 400	—	4 545	—	—	1 475	2 160	—	—	—
Total volatile solids	4 900	—	2 930	—	—	814	983	—	—	—
Suspended solids										
Average	—	4 277	2 205	1 518	7 772	383	143	740	1 075	4 408
Maximum	—	17 300	—	—	32 400	—	—	—	—	14 650
Minimum	—	660	—	—	100	—	—	—	—	100
Volatile suspended solids										
Average	—	2 645	1 660	—	4 403	299	118	—	750	3 176
Maximum	—	10 850	—	—	17 750	—	—	—	—	10 650
Minimum	—	420	—	—	60	—	—	—	—	75
Biochemical oxygen demand										
Average	—	713	—	—	1 238	—	—	—	515	667
Maximum	—	1 880	—	—	6 000	—	—	—	—	2 700
Minimum	—	200	—	—	135	—	—	—	—	100
Carbonaceous oxygen demand	—	—	4 565	2 230	—	1 384	1 310	1 230	—	—
Total organic carbon	—	—	1 242	—	—	443	320	—	—	—
Total (PO₄)–P	—	—	143	85	—	63	87	100	—	—
NH₃–N	—	—	853	—	—	253	559	—	480	—
Organic nitrogen	—	—	291	678	—	53	91	360	560	—
pH	8.0	—	7.3	7.2	—	7.0	7.8	7.0	7.3	—
Volatile acids	—	—	264	—	—	322	250	—	—	—
Alkalinity (as calcium carbonate)	2 555	—	3 780	—	—	1 349	1 434	—	—	—
Phenols										
Average	—	0.23	—	—	0.23	—	—	—	—	0.35
Maximum	—	0.80	—	—	0.50	—	—	—	—	1.00
Minimum	—	0.06	—	—	0.06	—	—	—	—	0.08

[a] Unless noted, all values are average for the sampling period studied.

[b] Values indicated are a composite from seven treatment plants.

[c] Values indicated are a composite from six treatment plants.

these alternatives the engineer must decide which pollutants must be removed and should determine the most economical method of treating the supernatant. This should then be compared with the costs associated with modifying the solids treatment train and expanding the treatment train to accommodate the supernatant.

ALKALINITY AND pH CONTROL. The methanogens present in the anaerobic digestion process are affected by even small pH changes, while the acid producers can function satisfactorily in a wide range of pH values. The effective pH range for methane producers is approximately 6.5 to 7.5 with an optimum range of 6.8 to 7.2. Maintenance of this optimum range is important to ensure effective gas production and to eliminate digester upsets.

The stability of the digestion process depends on the buffering capacity of the digester contents (that is, the ability of the digester contents to resist pH changes). Higher alkalinity values indicate a greater capacity for resisting pH changes. The alkalinity, which is more important in anaerobic digestion, is measured as bicarbonate alkalinity. Values for alkalinity in anaerobic digesters range from 1 500 to 5 000 mg/L as calcium carbonate. The volatile acids produced by the acid producers tend to depress pH. Volatile acid concentrations under stable conditions range from 50 to 100 mg/L. By maintaining a constant ratio of volatile acids to alkalinity below 0.25, the buffering capacity of the system can be maintained.

The bicarbonate alkalinity concentration can be calculated from the total alkalinity, which also comprises alkalinity of the volatile acids (such as acetate) and ammonium by the following formula:

$$\text{Bicarbonate alkalinity} = \text{Total alkalinity} - (0.71 \times \text{Volatile acids}) \quad (4.22)$$

Where

Bicarbonate alkalinity	=	mg/L as $CaCO_3$,
Total alkalinity	=	mg/L as $CaCO_3$,
Volatile acids	=	mg/L as acetic acid, and
0.71	=	conversion factor to mg/L as $CaCO_3$.

Barber and Dale (1978) have developed an equation to predict the necessary amount of bicarbonate alkalinity needed to raise the total alkalinity:

$$D_d = D_{\max}\left(1 - \frac{1}{\theta}\right) \quad (4.23)$$

Where

D_d	=	amount added daily to reach set level, mg/L as $CaCO_3$;
$D_{\max}$	=	required increase, mg/L as $CaCO_3$; and
$\dfrac{1}{\theta}$	=	reciprocal of average detention time or SRT, days^{-1}.

Sodium bicarbonate, lime, sodium carbonate, and ammonium hydroxide have been used successfully to increase the alkalinity of digester contents. However, most well-designed and well-operated digester facilities do not require alkalinity addition provided the wastewater has sufficient buffering capacity. The alkalinity of the wastewater should be evaluated before determining the need for building an alkalinity feed system.

STRUVITE. The formation of struvite, or magnesium ammonium phosphate ($MgNH_4PO_4$), scale deposits in digestion systems is a phenomenon thought to be common but has seldom been formally documented. Strictly speaking, it does not impair the anaerobic digestion process, but causes maintenance problems such as clogged pipes and heat exchangers, which are difficult to mitigate. This phenomenon is largely caused by the environmental factors unique to anaerobic digestion.

Chemical Aspects. The precipitation of magnesium ammonium phosphate is related to the chemistry of anaerobic digestion. The formation of ions during digestion is one aspect to be considered. Ammonium ions enter the system both as ions already in solution and through conversion of organic nitrogen to ammonia during anaerobic digestion. Phosphate ions likewise enter the system as both ions in solution and through the release of phosphate from organic forms during anaerobic digestion. (The release of phosphate from secondary biomass under anaerobic conditions is a familiar process used for the biological removal of phosphorus from wastewater.) Unlike ammonium and phosphate ions, however, practically all magnesium ions within the digester enter as solvated ions.

Because the release of ammonium and phosphate is continuous, the digesting biomass can become saturated, and possibly supersaturated, with magnesium ammonium phosphate. The concentrations of ammonium and phosphate ions can increase until, combined with the available magnesium concentration, they exceed the conditional solubility product of magnesium ammonium phosphate at the temperature and pH conditions in the digester.

The concept of "conditional solubility" refers to the specific set of actual conditions under which a given metal–salt compound may or may not be soluble (Stumm and Morgan, 1981). This is more appropriate than the "simpler" solubility product that only considers the product of all participating ion concentrations without regard to other equilibria in solution. Because many side reactions and the actual temperature and pH conditions can influence solubility behavior of a complex compound such as magnesium ammonium phosphate, conditional solubility is a more useful method of analysis.

Control Methods. Most alternatives that are used to deal with struvite scaling are preventive, such as smooth-lined piping and increased maintenance. Control methods that would attack the source of struvite precipitation would include such measures as dilution to reduce ion concentrations, and operating at a lower pH, neither of which is a desirable operation procedure.

Innovative research is needed to better control and reduce struvite scale formation in digestion systems. One promising method is the addition of an iron compound to precipitate phosphorus. Ferrous chloride was added to a California digestion system to control hydrogen sulfide generation, and a side benefit was the adjustment of the conditional solubility of magnesium ammonium phosphate to a level where the compound would be soluble. This apparently resulted from selective precipitation of the phosphorus by the ferrous compound. This observation suggests the concept of a "sacrificial digester" that could be operated solely to precipitate phosphorus, or even to precipitate struvite, before feeding the digestion system. Because phosphorus is an essential nutrient for the bacteria, care must be taken to ensure that sufficient phosphorus remains to meet nutritional needs.

After struvite deposits have formed, they are difficult to remove. Acid washing is an effective method, but can be costly and presents safety hazards. During early stages of formation, struvite may be controlled by frequent cleaning (pigging) of pipelines. Smooth-lined piping using polyvinyl chloride or glass-lined materials and polyethylene or polytetrafluoroethylene-coated plug valves will resist struvite accumulation better than other materials.

PERFORMANCE EVALUATION. Performance evaluation involves analysis and interpretation of data collected on critical performance indicators identified in the "Process Monitoring" section. Performance indicators are valuable because they provide early warnings of impending failure. Digester instability and failure can be initiated by changes in hydraulic, organic, or toxic loading; feed composition; temperature, or combinations of these. Such instability typically manifests itself as an increase in digester volatile acid concentration, and decreases in bicarbonate alkalinity, pH, gas production, gas methane content, and volatile solids destruction. Close monitoring of these indicators can enable the operator to initiate remedial measures in time to prevent system failure. Because of the complexity of process interactions, however, "measurement" of only a single variable will not yield an adequate description of process performance. For reliable monitoring, all variables should be measured frequently (Dague, 1968).

As a reference basis for recognition of an abnormal condition, Figure 4.22 defines an approximate alkalinity–pH–percent carbon dioxide "operating envelope" for normal anaerobic digestion (McCarty, 1964). Data outside this envelope should be considered indicative of upset conditions. Similarly, based on 0.351 m^3 methane produced per kg of COD destroyed (5.62 scf/lb) and a system mass balance, the typical methane content for municipal digesters is from 60 to 70% (Dague, 1968; Metcalf and Eddy, Inc., 1979; and Woods and Malina, 1965). Methane production may be the most sensitive performance indicator because it is directly related to COD destruction (Dague, 1968), but it is also affected by daily loading variations that do not lead to failure. It should, therefore, be used in conjunction with other indicators. Hydrogen gas plays a key role in anaerobic digestion and may also be used as a performance indicator. Typical BOD and total solids reductions for all types of feed are from 60 to 90% and from 45 to 50%, respectively. Biochemical oxygen demand and volatile solids reductions are nearly in the same range, and are reported from 40 to 70% for primary and from 20 to 50% for waste-activated sludges. Although BOD, COD, or volatile solids destruction are less sensitive indicators of process upsets, they can help determine the cause of the upset.

END-USE/DISPOSAL CONSIDERATIONS. End-use/disposal options for anaerobically stabilized biosolids include the following:

- Dewatering–drying–incineration;
- Land application, wet;
- Dewatering–land application;
- Land application, dedicated;
- Landfill, biosolids only;
- Landfill, codisposal; and
- Composting.

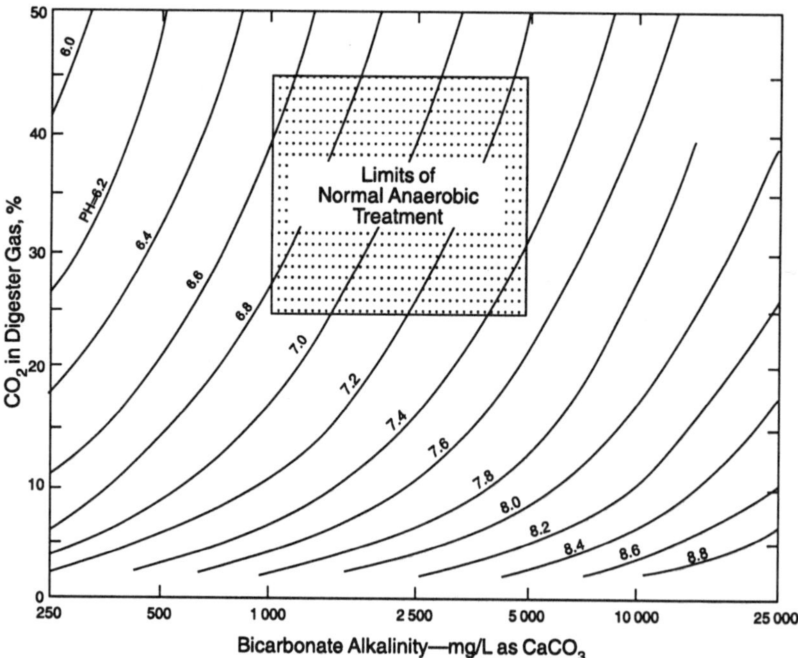

Figure 4.22 Relationship between bicarbonate alkalinity, pH, and carbon dioxide percentage in anaerobic treatment (35°C) (McCarty, 1964).

Criteria that determine the fate of a specific digested biosolids product are degree of stabilization (BOD/COD content), dewaterability, percent solids, nutrient content, metal content, pathogen content, odor potential, vector attraction potential, and economics.

Incineration is suitable for easily dewaterable biosolids with high organic content, whereas wet application can eliminate the dewatering step for hard-to-dewater biosolids. Pathogen, nutrient, vector attraction potential, and metal content also determine suitability for land application, dedicated or otherwise. Landfill disposal requires dewatering to achieve a high solids content (that is, greater than 20% dry weight). Odor potential is a consideration for all options. Anaerobic digestion typically produces a Class B biosolids as defined in the U.S. EPA's Final Rule *Standards for the Use or Disposal of Sewage Sludge* regulations and is suitable for restricted beneficial reuse on agricultural or forest land in any form (U.S. EPA, 1993).

P*ROCESS VARIATIONS*

TWO-PHASE ANAEROBIC DIGESTION. Two-phase digestion refers to the separation of the two major anaerobic reactions, acid formation and methane generation, to benefit the overall stabilization process. The most practical manner of achieving phase separation is kinetic control by regulating detention time and recycle ratio for each reactor (U.S. EPA, 1979). Figure 4.23 illustrates the major components of this system. Raw and recycle solids are blended, heated, and fed to a 1- to 2-day detention time vessel. This first vessel

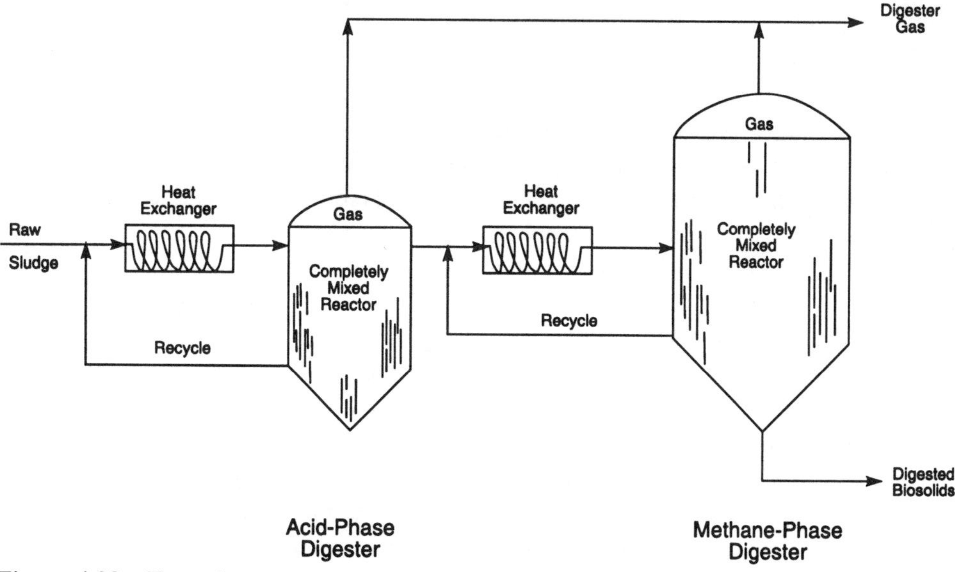

Figure 4.23   Two-phase anaerobic digestion process.

is referred to as the acid-phase digester. Hydrolysis of the suspended organic matter and subsequent formation of low molecular weight fatty acids is accomplished in this reactor. A low pH environment (5.5 to 6.5) is established and methane generation is negligible in this phase. This first phase can be operated at either mesophilic or thermophilic temperature ranges. Acid-phase biomass is then fed to a second vessel (methane-phase digester) providing approximately a 10-day detention time and typically operated at mesophilic range temperatures. Conditions in this second phase are similar to those found in conventional high-rate systems, which are operated to maintain an optimum environment for methanogenic bacteria.

Most of what is known about two-phase anaerobic digestion has been derived from small-scale pilot studies (Ghosh *et al.*, 1975; Ghosh *et al.*, 1987; and Kun *et al.*, 1989) and from data collected from two full-scale facilities (Ghosh *et al.*, 1991, and Peterson, 1990). This experience indicates that anaerobic digester performance can be improved by optimizing separately the acid-forming and methane-generation phases of the process. Advantages over single-phase systems cited include higher rates of volatile solids reduction and increased production rate and methane content of the final product gases, higher pathogen reduction capabilities, elimination of foaming problems, and an overall enhanced resilience and stability of the digestion process.

THERMOPHILIC ANAEROBIC DIGESTION. Most anaerobic digesters are operated in the mesophilic temperature range (35°C, or 95°F). They can, however, be operated in the thermophilic temperature range (55°C, or 130°F). Design criteria and system performance for thermophilic digestion are somewhat different from those for mesophilic digestion. Thermophilic digestion is thought to achieve higher volatile solids reduction per unit volume of digester than mesophilic digestion. Because the temperature is higher, however, the heating costs will be higher for thermophilic digestion.

In spite of these additional costs, thermophilic digestion is reported to have a number of advantages over mesophilic digestion. One reported advantage is that the dewatering characteristics of thermophilically digested biosolids are superior to those of mesophilically digested biosolids. Few data fully evaluate these claims. The possibility exists, however, that the additional costs of thermophilic digestion may be offset by reduced dewatering costs.

A second potential advantage of thermophilic digestion is that there may be a greater reduction in pathogen levels than for mesophilically digested biosolids. However, the most recent federal regulations controlling land application of biosolids classify both mesophilic and thermophilic anaerobic digestion as a process to significantly reduce pathogens, rather than classifying thermophilic digestion as a process to further reduce pathogens. Therefore, no credit is given to thermophilic anaerobic digestion for the potentially greater reduction in pathogens.

Thermophilic digesters are more difficult to operate than mesophilic digesters because they are more sensitive to rapid changes in temperature. Two-stage digestion consisting of a mesophilic digester followed by a thermophilic digester has been used to solve this problem (Torpey *et al.*, 1984).

In summary, thermophilic digestion is more costly than mesophilic digestion. It has been claimed that these increased costs can be offset by decreased dewatering costs and increased reduction of the pathogen levels of these biosolids. To date, these claims have not been completely verified. Pilot testing with the actual feed is recommended before deciding to use thermophilic digestion.

*R*EFERENCES

Anthony, R.M., and Breimhurst, L.H. (1981) Determining maximum influent concentrations of priority pollutants for treatment plants. *J. Water Pollut. Control Fed.,* **53,** 1457.

Barber, N.R., and Dale, C.W. (1978) Increasing Sludge Digester Efficiency. *Chem. Eng.*

Benefield, L.D., and Randall, C.W. (1980) *Biological Process Design for Wastewater Treatment.* Prentice–Hall, Inc., Englewood Cliffs, N.J.

Burd, R.S. (1968) A Study of Sludge Handling and Disposal. Pub. WP-20-4, Fed. Water Pollut. Control Admin. Dep. Interior, Washington, D.C.

Buswell, A.M., and Neave, S.L. (1939) Laboratory Studies of Sludge Digestion. Ill. State Water Surv. Bull., **30**.

Camp, T.R., and Stein, P.C. (1943) Velocity Gradients and Internal Work in Fluid Motion. *J. Boston Soc. Civ. Eng.,* **203**.

Dague, R.R. (1968) Application of digestion theory to digester control. *J. Water Pollut. Control Fed.,* **40,** 2021.

DeWalle, F.B., *et al.* (1979) Heavy metal removal with completely mixed anaerobic filter. *J. Water Pollut. Control Fed.,* **51,** 22.

Eastman, J.A., and Ferguson, J.F. (1981) Solubilization of particulate organic carbon during the acid phase of anaerobic digestion. *J. Water Pollut. Control Fed.,* **53,** 352.

Farell, J.B., *et al.* (1986) Microbial Destructions Achieved by Full-scale Anaerobic Digestion. EPA-600/D-86-1031, U.S. EPA, Cincinnati, Ohio.

Garber, W.F. (1982) Operating experience with thermophilic anaerobic digestion. *J. Water Pollut. Control Fed.,* **54,** 1170.

Ghosh, S., *et al.* (1975) Anaerobic acidogenesis of wastewater sludge. *J. Water Pollut. Control Fed.,* **47,** 30.

Ghosh, S., *et al.* (1987) *Stabilization of Sewage Sludge by Two-Phase Anaerobic Digestion.* EPA-7600/2-87-040, U.S. EPA, Cincinnati, Ohio.

Ghosh, S., *et al.* (1991) Pilot and Full-Scale Studies on Two-Phase Anaerobic Digestion for Improved Sludge Stabilization and Foam Control. Paper presented at 64th Annu. Conf. Water Pollut. Control Fed., Toronto, Can.

Haug, R.T., *et al.* (1978) Effects of thermal pretreatment on digestibility and dewaterability of organic sludges. *J. Water Pollut. Control Fed.,* **50,** 73.

Health Education Services Inc. (1978) *Recommended Standards for Sewage Works.* Albany, N.Y.

Kroker, E.J., *et al.* (1979) Anaerobic treatment process stability. *J. Water Pollut. Control Fed.,* **51,** 718.

Kugelman, I.J., and McCarty, P.L. (1965) Cation Toxicity and Stimulation in Anaerobic Waste Treatment, Part II—Daily Feed Studies. *Proc. 19th Ind. Waste Conf., Purdue Univ., Ext. Ser. 1965,* West Lafayette, Ind.

Kun, M.L., *et al.* (1989) Destruction of enteric bacteria and virus during two-phase digestion. *J. Water Pollut. Control Fed.,* **61,** 1421.

Lawrence, A.W. (1971) Application of Process Kinetics to Design of Anaerobic Processes. In *Anaerobic Biological Treatment Processes.* F.G. Pohland (Ed.), Am. Chem. Soc. Adv. Chem. Ser., 105.

Lawrence, A.W., and McCarty, P.L. (1965) The role of sulfide in preventing heavy metal toxicity in anaerobic treatment. *J. Water Pollut. Control Fed.,* **37,** 392.

Lawrence, A.W., and McCarty, P.L. (1974) A Unified Basis for Biological Treatment Design and Operation. *J. Sanit. Eng. Div., Proc. Am. Soc. Civ. Eng.* **96,** 757.

Liptak, B.G. (1974) *Environmental Engineers' Handbook.* Chilton Book Co., Radnor, Pa.

McCarty, P.L. (1964) Anaerobic Waste Treatment Fundamentals; I: Chemistry and Microbiology; II: Environmental Requirements and Control; III: Toxic Materials and Their Control; IV: Process Design. *Public Works,* 9, 10, 11, and 12.

Metcalf and Eddy, Inc. (1979) *Wastewater Engineering: Treatment, Disposal, Reuse.* 2nd Ed., G. Tchobanoglous (Ed.), McGraw–Hill, Inc., New York, N.Y.

Mosey, F.E. (1976) Assessment of the Maximum Concentration of Heavy Metals in Crude Sludge Which Will Not Inhibit the Anaerobic Digestion of Sludge. *Water Pollut. Control* (G.B.), **75,** 10.

National Fire Protection Association (1993) National Electric Code. Quincy, Mass.

O'Rourke, J.T. (1968) Kinetics of Anaerobic Treatment at Reduced Temperatures. Ph.D. thesis, Stanford Univ., Palo Alto, Calif.

Peterson, A.C. (1990) Kansas City's Full-Scale Two-phase Digestion System. Paper presented at Energy From Biomass Waste Symp.

Rudolphs, R.H., and Fontenelli, L.S. (1945) Supernatant liquor treatment with chemicals. *Sew. Works J.,* **17,** 538.

Speece, R.E. (1972) Anaerobic Treatment. In *Process Design in Water Quality Engineering.* E.L. Thackston and W.W. Eckenfelder (Eds.), Jenkins Publishing Company, N.Y.

Storey, G.W. (1987) Survival of Tapeworm Eggs, Free and in Progottids, During Simulated Sewage Treatment Processes. *Water Res.* (G.B.), **2,** 199.

Stukenberg, J.R., *et al.* (1990) Egg-Shaped Digesters: From Germany to the United States. Paper Presented at 63rd Annu. Conf. Water Pollut. Control Fed., Washington, D.C.

Stukenberg, J.R., *et al.* (1992) Compliance Outlook: Meeting the 40 CFR Part 50 Pathogen Reduction Criteria with Anaerobic Digestion. Paper Presented at 65th Annu. Conf. Water Environ. Fed., New Orleans, La.

Stumm, W., and Morgan, J.J. (1981) *Aquatic Chemistry.* 2nd Ed., John Wiley and Sons, New York, N.Y.

Torpey, W.N., *et al.* (1984) Effects of multiple digestion on sludge. *J. Water Pollut. Control Fed.,* **56,** 62.

Tortorici, L., and Stahl, J.F. (1977) Waste Activated Sludge Research. *Proc. Sludge Manage., Disposal, Utilization Conf.,* Inf. Transfer Inc., Rockville, Md.

U.S. Environmental Protection Agency (1979) *Process Design Manual, Sludge Treatment and Disposal.* Munic. Environ. Res. Lab. Cincinnati, Ohio.

U.S. Environmental Protection Agency (1993) *Standards for the Use or Disposal of Sewage Sludge. Fed. Regist.,* **58,** 32, 40 CFR Part 503.

Water Environment Federation (1992) *Design of Municipal Wastewater Treatment Plants.* Manual of Practice No. 8, Alexandria, Va.; Am. Soc. Civ. Eng., Manual and Report on Engineering Practice No. 76, New York, N.Y.

Water Environment Federation (1994) *Safety and Health in Wastewater Systems.* Manual of Practice No. 1, Alexandria, Va.

Water Pollution Control Federation (1980) *Sludge Thickening.* Manual of Practice No. FD-1, Washington, D.C.

Water Pollution Control Federation (1987) *Anaerobic Sludge Digestion.* Manual of Practice No. 16, Alexandria, Va.

Woods, C.E., and Malina, V.F. (1965) Stage digestion of wastewater sludges. *J. Water Pollut. Control Fed.,* **37,** 1495.

Chapter 5
Composting

Composting is a biochemical method of stabilization that prepares wastewater residuals for beneficial use as a soil conditioner. It is a self-heating process that destroys pathogens and produces a material similar to soil humus. Well-stabilized compost can be stored indefinitely and has minimal odor even if rewetted. Compost is suitable for a variety of end uses such as landscaping, topsoil blending, potting, and growth media, and may be distributed to the general public for gardening. It may also be used in agriculture for soil erosion control, improvement of soil physical properties, and in revegetation of disturbed lands. Local markets may be developed in urban and nonagricultural areas as well as in agriculture and mine revegetation.

During composting a medium dry enough to provide pore spaces with free air, but wet enough to sustain biological activity, typically 35 to 50% solids, is used. Porosity typically is provided by mixing dewatered cake with a bulking agent or amendment, such as wood chips. Because of the bulking agent, the volume of compost product is equal to or greater than the volume of dewatered cake. For a given amount of dry solids, the volume of material to be composted increases with decreasing percent solids because of the greater amount of moisture.

Raw, digested, or chemically stabilized solids may be composted. The process has also been proven effective for organic residuals from the paper, pharmaceutical, and food-processing industries. The bulking agent or amendment may be a wide variety of materials including other wastes such as wood wastes, yard waste, paper, agricultural residue, wood ash, and animal bedding.

Composting is a relatively simple process that can be performed outdoors in most climates. Because of a desire to operate the process more efficiently, control odors, and reduce operating costs, many facilities are constructed under structures, in fully enclosed buildings, or in entirely mechanized facilities.

This chapter includes a summary of the basic theory of composting, design considerations, a description of the most commonly used composting methods, the source of odors generated during composting and methods of their control and treatment, the general principals of composting operations, and compost product use and marketing.

*P*ROCESS DESCRIPTION

Figure 5.1 is a generalized flow chart applicable to most composting processes. It applies to the three basic types of composting system:

- Aerated static pile—dewatered cake is mixed with a coarse bulking agent such as wood chips, and the mixture is stacked over a porous bed with air piping connected to blowers. The piles are covered with a layer of finished compost to provide insulation and capture odor. Air is drawn downward or forced upward through the mixture. After composting, the pile is taken down. The bulking agent may be partially recovered by screening and reused.
- Windrow process—the mixture is stacked in windrows (long narrow piles) with a sufficient ratio of surface area to volume to provide aeration by natural convection and diffusion. The windrow is remixed

periodically by a turning machine. The amendments typically are of smaller particle size than with aerated static pile and may include recirculated compost. In the aerated windrow process, natural convection and diffusion are supplemented by forced aeration, as in the static pile process. Air is supplied through trenches in the paved working surface.

- In-vessel processes—the mixture is fed into one end of a silo, tunnel, or open channel and moves continuously toward the discharge end, where it is outloaded after the required detention time. Air is forced through the mixture. The mixture may move as an undisturbed plug or be periodically agitated as it is moved through the vessel.

In all the process variations, the bulking agent or amendment increases porosity so the mixture can be aerated and also decreases the percentage moisture content of the mixture. A bulking agent consists of coarse particles to increase porosity. An amendment may consist of smaller particles. Either may provide supplemental available carbon to adjust the energy balance and carbon-to-nitrogen ratio. These concepts are discussed in more detail in a later section.

The microbial activity in the composting step requires oxygen and generates carbon dioxide, water vapor, and heat. The temperature of the mixture may exceed 70°C (158°F), although the optimum range is 50 to 60°C (122 to 140°F). After 3 to 10 days, the temperature gradually decreases. In addition to supplying oxygen, aeration or agitation removes the exhaust gases, water vapor, and heat. The rate of aeration may be used to control the process temperature and rate of drying.

The rapidly biodegradable organic materials are converted through a series of metabolic processes into more stable materials. This process continues at a much lower rate in the curing step. In curing, the oxygen demand and rate of heat generation typically are low enough that forced aeration or agitation are not essential if the material is sufficiently porous. However, there is a trend toward aerated curing to ensure aerobic conditions and control odors. The

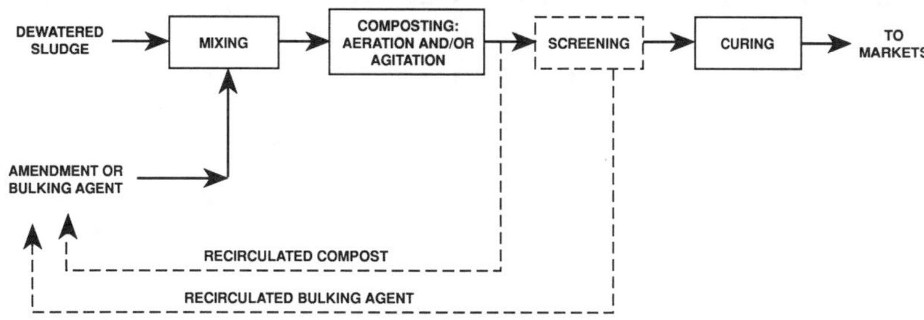

Figure 5.1 Generalized composting flow chart (dashed lines indicate optional steps; screening may follow curing; drying step may precede screening).

combined detention time for active composting and curing typically ranges from 50 to 80 days.

An optional screening step removes bulking agent for reuse. The screen improves product quality by removing woody material and large particles and reduces the need for new bulking agent.

MICROBIOLOGY. Three major categories of microorganisms involved in the composting process are bacteria, actinomycetes, and fungi. The bacteria are responsible for a major portion of organic matter decomposition. Initially, at mesophilic temperatures (lower than 40°C, or 104°F), they metabolize carbohydrates, sugars, and proteins. At thermophilic temperatures (higher than 40°C), they decompose protein, lipids, and the hemicellulose fractions. They are undoubtedly responsible for much of the heat energy produced.

The role of the actinomycetes (microorganisms common to soil environments) is not clear. Waksman and Cordon (1939) indicated that this group attacks hemicellulose, but not cellulose. They metabolize a wide variety of organic compounds, such as sugars, starches, lignin, proteins, organic acids, and polypeptides. Stutzenberger (1971) has isolated a thermophilic actinomycete that may be important in cellulose degradation.

The fungi are also present at both the mesophilic and thermophilic temperature ranges. Chang (1967) indicated that mesophilic fungi metabolize cellulose and other complex carbon sources. Their activity is similar to the actinomycetes. Both the fungi and actinomycetes typically are found in the exterior portions of compost piles. Golueke (1977) suggests that this phenomenon is related to the aerobic nature of the organisms because most of the fungi and actinomycetes are obligate aerobes.

Microbial activity during composting can be classified into three basic stages: mesophilic, when temperatures in the pile range from ambient to 40°C (104°F); thermophilic, when temperatures range from 40 to 70°C (104 to 158°F); and a cooling period associated with a reduction of microbial activity and completion of the composting process. The optimum temperature in the thermophilic range appears to be between 55 and 60°C (131 and 140°F), and results in the maximum rate of volatile solids destruction.

Biological solids, newly harvested wood waste, and yard waste provide a diverse population of microflora that can respond to changing temperatures and changes in the substrate. Under most circumstances, inoculum of pure cultures does not significantly enhance the composting process. Inoculation with a cellulose-decomposing fungus combined with nutrient addition has proven effective in accelerating sawdust decomposition. Thermal conditioning may sterilize the organic biomass.

ENERGY BALANCE. Heat is generated by the conversion of organic carbon to carbon dioxide and water vapor. The fuel is provided by the rapidly degraded volatile fraction of the solids supplemented by the amendment.

Heat is removed primarily by evaporative cooling through aeration and agitation. To a lesser extent, heat is lost at the pile surface. The process temperature will not rise if the rate of heat loss exceeds the rate of heat generation. Haug (1980) provides a detailed discussion of the energy balance, concluding with the following relationship:

$$W = \frac{\text{Weight of water evaporated (lb or kg)}}{\text{Loss of volatile solids (lb or kg)}} \qquad (5.1)$$

If W is below 8 to 10, sufficient energy should be available for heating and evaporation. If W exceeds 10, the mix will remain cool and wet. This generalization is based on heat of vaporization and does not consider the effect of ambient conditions on evaporation and surface cooling.

CARBON-TO-NITROGEN RATIO. Microorganisms use carbon and nitrogen in proportions fixed by the composition of the microbial biomass. The ideal ratio of available carbon to nitrogen (C:N) is in the range of 25:1 to 35:1. If the C:N ratio is less than 25:1, excess nitrogen will be released as ammonia, resulting in loss of nutrient value and emission of ammonia odor. If the C:N ratio exceeds 35:1, organic material will break down more slowly and remain active well into the curing stage (Poincelot, 1975). The C:N ratio of wastewater residuals typically is in the range of 5 to 20. The amendment and bulking agent provide supplemental carbon, improving both the energy balance and the C:N ratio of the mixture.

The calculation of C:N ratio is complicated by the fact that some of the carbon becomes available more slowly than the nitrogen (Kayhanian and Tchobanoglous, 1992). If wood chips are used as a bulking agent, only a thin surface layer of the wood provides available carbon. Carbon in sawdust is more rapidly available.

P*ROCESS OBJECTIVES*

The primary objective of composting is to produce a fertilizerlike product that can be beneficially used. The compost must meet regulatory and public health requirements and be attractive for some end use. This primary objective is met through the following process objectives:

- Pathogen reduction,
- Maturation, and
- Drying.

PATHOGEN REDUCTION. Pathogenic organisms found in wastewater residuals fall into five major groups: bacteria, viruses, protozoa cysts, helminthic (parasitic worm) ova, and fungi.

The first four groups are often termed primary pathogens because they can invade normally healthy persons and cause diseases. The last group, the fungi, are referred to as secondary pathogens because they typically only infect persons with weakened respiratory or immune systems.

Elevated temperature is one of the most effective means of destroying pathogens. Table 5.1 summarizes time/temperature relationships for pathogen inactivation in actual composting operations. It is important to note that temperatures measured within a composting pile or vessel may not be uniform because of variations in heat loss, mix characteristics, and air flow.

The Water Pollution Control Federation (1984) reported that composting where temperatures reach the thermophilic range should eliminate practically all viral, bacterial, and parasitic pathogens. However, some fungi, such as *Aspergillus fumigatus*, are thermotolerant and survive the composting process.

Data from Los Angeles, California, on bacterial concentrations in the windrows showed marked reduction within 15 days (Iacoboni *et al.*, 1980). At 20 days, no *Salmonella* were detected. Fecal and total coliforms survive during windrow composting in a cool humid climate. *Salmonella*, however, were eliminated after 14 days. Studies using an F_2 bacteriophage virus showed survival for as long as 45 days for digested biosolids and more than 55 days when undigested.

Data on static pile composting show that total coliforms, fecal coliforms, and *Salmonella* were not detected after 10 days of composting when temperatures exceeded 55°C (130°F) for several days. Further studies using an F_2 bacteriophage virus as an indicator of virus destruction revealed that with static pile composting, destruction was achieved in 14 days.

Data on parasite survival during composting are limited. Analysis of compost in Windsor, Ontario, and Puerto Rico did not show any viable ova.

Table 5.1 **Temperature and time required for pathogen destruction in compost (Knoll, 1964; Morgan and MacDonald, 1969; Shell and Boyd, 1969; and Wiley and Westerberg, 1969).**

	Exposure time for destruction at various temperatures, hours		
Microorganisms	**45–55°C**	**60°C**	**65°C**
Salmonella newport		25	
Salmonella	168	116	
Poliovirus type 1		1.0	
Candida albicans		72	
Ascaris lumbircoides		4.0	1.0
Mycobacterium tuberculosis			336

Wiley and Westerberg (1969) were unable to detect viable *Ascaris* ova 1 hour after seeding the ova into a high-rate composting system operating at 60 to 65°C (140 to 150°F). Yanko (1987) never recovered viable helminth ova in examining approximately 300 samples of compost although they were plentiful in the feed.

Salmonella can regrow in finished compost, but parasite ova and virus cannot. Regrowth can be reduced by avoiding handling finished compost with equipment contaminated with raw feed.

Many microorganisms can function as secondary pathogens, but composting conditions favor the growth of some more than others. Millner *et al.* (1977) report that the fungus *A. fumigatus* Fres (Af) has been isolated at relatively high concentrations from finished compost and from compost pile zones at less than 60°C (140°F). Other secondary fungal pathogens isolated occasionally from compost are *Mucor pusillus* and *M. miebei*. These fungi, common to composting operations, typically are found in back yards, decayed leaves, grass, commonly available organic soil amendments, and ventilation ducts.

During certain composting operations the number of *A. fumigatus* spores released to the atmosphere is increased. LeBrun (1979) reported on windrow and reactor studies at the Los Angeles County Sanitation District and found levels of 10^3 to 10^4 cfu (colony forming units)/g initially. These levels were reduced to 10 cfu/g at the end of the composting period.

Serratia marcescens and *Pseudomonas* spp. are secondary bacterial pathogens isolated from compost, but their heat sensitivity and inability to form spores seem to limit their occurrence. Conversely, some thermophilic actinomycetes proliferate during composting. These can incite respiratory allergic responses in certain sensitized people. Exposure to airborne spores is minimized by controlling dust. Compost should not be allowed to become too dry. Workers should be provided with dust masks when working in dusty areas.

MATURATION. Maturation is the conversion of the rapidly biodegradable components in the organic material and amendment into substances similar to soil humus, which decomposes very slowly. Compost that is insufficiently mature will reheat and generate odors in storage and upon rewetting. It may inhibit seed germination by generating organic acids and inhibit plant growth by removing nitrogen as it decomposes in the soil. Stability refers to the reduction in the rate of microbial degradation of the biodegradable components in the mixture as the compost matures.

There are a number of testing methods and standards for measuring for compost stability or maturity. At the present time, no single test is accepted universally (Jimenez and Garcia, 1989). The standards associated with each test are still tentative and much work needs to be done to correlate test results with odor generation and plant growth. A complete assessment of maturity may require more than one type of test.

Volatile solids as percentage of total solids is not a good measure of stability because it fails to account for the rate of biodegradation. Respiration tests measuring carbon dioxide production or oxygen demand are better representations of stability but are sensitive to the conditions of the test. Measurements of carbon dioxide production typically are performed directly on the compost mix in an incubator. They are useful simulations of the composting process and can be used on highly unstable samples from early in the process, as well as on finished compost. Oxygen uptake rates may be measured on the compost mix or in an aqueous extract. The latter is determined using the standard oxygen uptake rate test.

The C:N ratio in mature compost should be less than 20. Available carbon in the compost can deplete soil nitrogen, because nitrogen is required by soil microorganisms that use the newly introduced carbon. Seed germination and root elongation tests are performed by germinating seeds (such as cress) in a filtered extract of the compost and comparing with a control using distilled water. This test measures phytotoxicity caused by organic acids in the compost. Stabilization is achieved by maintaining optimal conditions for a sufficient period of time. Cellulosic materials, such as wood and yard waste, take more time to decompose than wastewater residuals; therefore, screening out of the bulking agent may improve stability.

DRYING. Water vapor is removed during composting, increasing the solids content of the mixture from approximately 40 to 55%. Drying is critical in processes that include screening, because screens do not perform well on compost with less than 50 to 55% solids. Drying is achieved by providing sufficient aeration or agitation to optimize the removal of water vapor.

*D*ESIGN CONSIDERATIONS

This section provides ranges of design parameters for each stage of the composting process and identifies the design criteria essential to successful operation.

BULKING AGENTS AND AMENDMENTS. The aerated static pile process and some in-vessel processes require mixes with sufficiently high porosity to be aerated with low-pressure blowers. For these processes, bulking agent is required. Table 5.2 lists some typically used bulking agents and their characteristics.

Bulking agents perform the following functions:

- Adjust the solids content of the mixture,
- Provide supplemental carbon to adjust the C:N ratio and energy balance, and

- Provide structural integrity to maintain porosity as the mixture is stacked.

Grass clippings typically are unsuitable as bulking agent because of their high water and nitrogen content and lack of porosity. If grass clippings are composted, they require supplemental bulking agent.

Windrow processes and some in-vessel processes do not require large pore spaces but still require adjustment of the solids content and supplemental carbon. In these cases, the material added is called the amendment. Sawdust and recirculated compost are frequently used as amendments.

Table 5.2 Characteristics of bulking agents.

Wood chips (1 – 2 in.)*	• Must typically be purchased • High recovery in screening (60–80%) • Good source of supplemental carbon
Chipped brush	• May be available as waste material • Low recovery rate in screening (40–60%) because of higher percentage of fines • Good source of supplemental carbon • Curing time may be prolonged because of continued breakdown of unrecovered fines
Ground waste lumber	• May be available as waste material • May be poor source of supplemental carbon if old and extremely dry, because more volatile forms of carbon are missing
Leaves	• Must be ground • Wide range of moisture content requires close attention to materials balance • Rapidly available source of supplemental carbon, resulting in high temperatures • Relatively low porosity • Cannot be recovered by screening, resulting in large volume of compost
Shredded tires	• Typically used in mix with other bulking agents • Provides no supplemental carbon • Almost 100% recoverable • May introduce metals
Shredded paper	• Typically only suitable in agitated bed or windrow systems • Slow to break down and may require long cure time

* in. × 25.40 = mm.

The following mixing technologies are used in composting operations:

- Front-end loader—the mix quality typically is poor and mixing is time consuming. Loaders are appropriate only for small facilities and emergencies.
- Batch mixer—mobile or stationary self-unloading batch mixers are well suited for small- and medium-size facilities. For each batch the correct proportions of dewatered cake and bulking agent are loaded in and mixed by internal flights.
- Continuous mixer—dewatered cake and bulking agents are fed continuously to the mixer from live bottom bins and proportioned automatically. Pug mills and plough mixers are examples of commonly used continuous mixers. This system is more complex than the batch mixer but well suited to medium- and large-size facilities.
- Windrow turner—materials are layered and mixed on a pad with a mobile turning machine. This technology is most suitable for windrow composting operations. The agitation in a horizontal agitated bed reactor also provides mixing, but premixing is recommended before loading material into the reactor.

MIX CHARACTERISTICS. The initial solids content of the mixture should be in the range of 40 to 45%. A wet mixture may have insufficient porosity and an unfavorable energy balance. Agitated bed, in-vessel systems are sometimes designed for mixtures in the range of 35 to 40% solids, because the daily agitation restores porosity and promotes evaporation.

Table 5.3 shows the ratios of bulking agent to dewatered cake required to achieve a mix solids content of 40%. The method of mixing is as important as the mix ratio. It is critical that the mix have uniform porosity and that all particles of cake be in close contact with the bulking agent. Dewatered cake of 18 to 25% solids should be mixed with bulking agent so that each wood chip or other bulking agent particle is coated with a thin layer of sludge. Dewatered cake of 30 to 35% solids will break into clumps. These clumps must be uniformly small and mixed with the bulking agent. Large clumps and balls will not compost and will contribute to odors. If mixing is not uniform, zones with a disproportionate amount of bulking agent will divert the flow of air, allowing other zones to become anaerobic.

CALCULATION OF MATERIALS BALANCE. Materials balance calculations track the weight and volume through each stage of the composting process. Table 5.4 shows a typical materials balance for an aerated static pile process using anaerobically digested biosolids dewatered to 20% solids. The biosolids are mixed with wood chips, stacked over a layer of wood chips to provide air distribution, and covered with a layer of unscreened compost. A typical pile cross section is shown in Figure 5.2. The entire pile is screened after composting and the oversized fraction is recycled as bulking agent.

A volume equal to the cover layer is set aside and not screened. The screen recovers 65% of the total input bulking agent by volume so it must be supplemented with makeup chips. In practice, the recovery rate is highly sensitive to moisture content or stickiness of the compost, percent fines in the bulking agent, and screen loading rate, and will be in the range of 50 to 80% by volume. The following input assumptions are required:

- The density of each material,
- The volatile solids reduction of each input, and
- The recovery efficiency of the screen.

Table 5.3   Calculation of ratio of bulking agent to dewatered cake.[a]

Cake solids	Bulking agent solids, %	Mix ratio, cu yd bulking agent/cu yd cake
As function of cake solids		
16	55	3.30
18	55	3.02
20	55	2.75
22	55	2.47
24	55	2.20
26	55	1.92
As function of bulking agent solids (cake at 20%)		
20	45	8.33
20	50	4.17
20	55	2.75
20	60	2.08

Assumptions:
Mix solids content—40%
Bulk density—1 650 lb/cu yd[b]
Bulking agent density—800 lb/cu yd[b]

[a] This table applies to static pile processes using wood chips. Extrapolation to higher solids content may lead to insufficient bulking agent to provide porosity.
[b] lb/cu yd $\times$ 0.5933 = kg/m^3.

Table 5.4   Example of materials balance for aerated static pile composting.

Item	Volume, cu yd[a] (A)	Total, lb[b] (B)	Dry wt, lb (C)	Volatile solids, dry lb (D)	Bulk solids density, lb/cu ft[c] (E)	Total Content, % (F)	Total Volatile, % (G)
(1) Dewatered cake	6.25	10 000	2 000	1 400	1 600	20	70
(2) Makeup chips	8	4 000	2 400	2 280	500	60	95
(3) Recycle chips	11	8 800	4 840	4 114	800	55	85
(4) Mix	20.7	22 800	9 240	7 794	1 100	40.5	84.3
(5) Base (recycle chip)	2.6	1 300	780	663	500	60	85
(6) Cover (unscreened compost)	2.6	2 080	1 144	961	800	55	84
(7) Total		26 180	11 164	9 418			
(8) Volatile solids loss			882	882			
(9) Unscreened compost	23.4	18 695	10 282	8 536	800	55	83
(10) Recovered chips	14.0	11 200	6 160	5 236	800	55	85
(11) Compost	7.5	7 495	4 122	3 300	1 000	55	80

Assumptions:

Volatile solids loss:

   Sludge—20 percentage points

   Makeup chips—10 percentage points

   Recycle chips—5 percentage points

Recovery by screen—65%

Pile height—8 ft of mix, 1 ft base, 1 ft cover (ft × 0.304 8 = m).

[a] cu yd × 0.764 6 = m^3.

[b] lb × 0.453 6 = kg.

[c] lb cu ft × 16.02 = kg/m^3.

Table 5.4 is constructed according to the following sequence of inputs and calculations:

Dewatered cake (1):	Begin with 1 dry ton or 2 000 lb dry weight. (C) (F) (G) and (H) must be known. (A) (B) and (D) are then calculated.
Makeup chips (2):	(F) (G) and (H) must be known. The target solids content in the mix (4F) is 40% (calculation shows 40.5%). The quantity of makeup chips is determined by this target.
Recycle chips (3):	Must equal recovered chips minus base, based on known or assumed recovery rate. For (9) the assumed solids content is 55% because it is difficult to screen at a lower solids content. This is a target.
Mix (4):	Columns (B) (C) and (D) are summed for rows (1) (2) and (3). (F) and (G) are back calculated from (B) (C) and (D). (F) must be known or assumed. This is critical in calculating the physical size of a facility and material handling requirements. (A) is back calculated from (F). Column (A) is not summed as volume is not additive.
Base (5):	Same characteristics as recycle chips. Volume (A) is a percentage of volume of mix (4A) depending on pile geometry. Example is based on 8-ft mix height, 1-ft base.
Cover (6):	Same characteristics as compost entering screen (8). Volume is a percentage of volume of mix depending on pile geometry. Example is based on 8-ft mix height 1-ft cover.
Total (7):	Columns (B) (C) and (D) are summed over rows (4) (5) and (6).
Volatile solids loss (8):	The percentage point drop in volatile solids must be known or assumed, for dewatered cake and wood chips. (8D) is then calculated by subtraction. (8C) = (8D).
Unscreened compost (9):	(C9) = (C7)–(C8). (D9) = (D7)–C(D8). Solids content (F9) is a target value and is assumed. (E9) is assumed. (B9) and (A9) are then calculated.
Recovered chips (10):	Based on assumed volumetric recovery rate. In this example, 65% (10A) = 65% × [(2A)+(3A)+(5A)].
Compost (11):	(11B) = (9B)–(10B). Density is assumed or known. Other entries can be calculated.

In formulating the mix, the target solids content of the mix should be 40% or more. This example shows 40.5%. The output solids content is assumed to be 55%, because that is the optimum solids content for efficient screening. Wetter compost will blind the screen and stick to the wood chips. Drier compost will create dust.

Because much of the bulking agent enters the compost, there is a slight gain in volume and percent volatile solids as compared with the dewatered cake input. Unlike with anaerobic digestion, the degree of stabilization cannot be measured as volatile solids reduction. However, the compost is stabilized as defined by reduction in respiration rate and odor generation.

The materials balance may be used to check the energy balance. As stated previously, the ratio of water evaporated-to-volatile solids loss should not exceed 10. In this materials balance

$$\frac{\text{Water evaporated}}{\text{Volatile solids loss}} = \frac{6\ 603\ \text{lb}}{882\ \text{lb}} = 7.48 \tag{5.2}$$

Therefore, this material balance is realistic with regard to energy availability.

TEMPERATURE CONTROL AND AERATION. Aeration removes heat and water vapor and supplies oxygen to the microorganisms. As the rate of air flow is increased in a forced aeration system, the pile temperature decreases and the rate of water vapor removal increases. Agitation releases heat and water vapor quickly and also enhances aeration by improving porosity. Without sufficient aeration, the pile temperature may exceed 70°C (158°F), which is detrimental to microbial activity. The optimum temperature range for volatile solids destruction is 40 to 50°C (approximately 105 to 120°F). In addition, 40 to 50°C is optimal for removal of water vapor. This is true because high rates of air flow are required to maintain the low temperatures in a

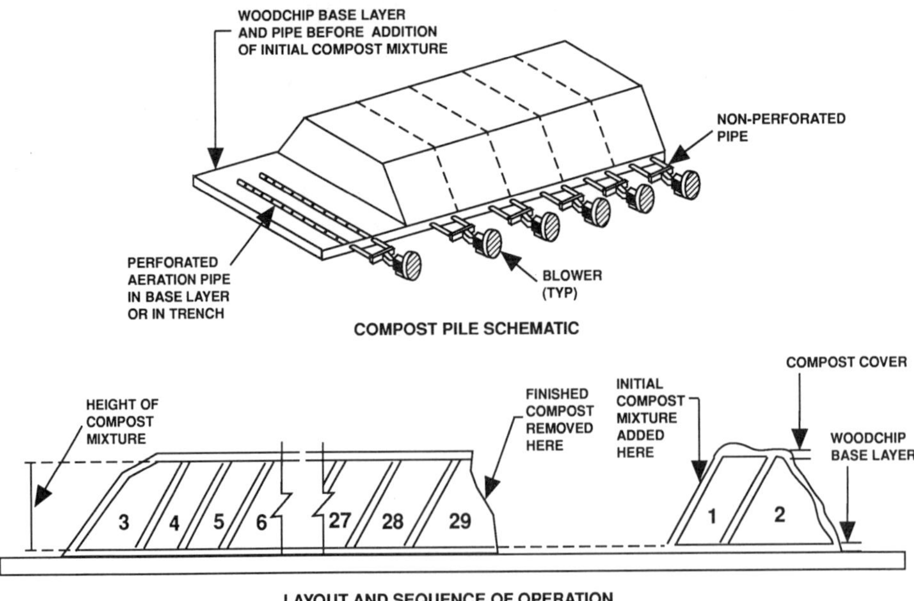

Figure 5.2   Aerated static pile composting.

highly active process. To ensure pathogen destruction, the temperature should be more than 55°C for part of the composting period.

Higgins *et al.* (1982) reported that an aeration rate of 34 m^3/Mg·h (1 100 cu ft/hr/dry ton) provided adequate drying and high enough temperatures for pathogen destruction. Early in the composting process, higher aeration rates may be needed to prevent excessive temperature.

To maintain temperatures less than 60°C during peak activity may require aeration rates approaching 300 m^3/Mg·h (10 000 cu ft/hr/dry ton). This aeration capacity may be impractical in large systems. Practical aeration capacities are in the range of 90 to 160 m^3/Mg·h (3 000 to 5 000 cu ft/hr/dry ton). Aeration in this range will control temperatures throughout most of the composting period and provide adequate moisture removal capacity. Higher aeration rates are possible but require larger diameter and more closely spaced piping or trenches. If highly reactive bulking materials are used, such as ground up leaves, the mass of bulking agent and dewatered cake may enter into the sizing of aeration capacity. In negative aeration, actual air flow (m^3) will be higher than standard air flow because of the elevated temperature and humidity of the air. Near the end of the process, after pathogen destruction requirements are met, higher aeration rates may be helpful to increase drying.

In the aerated static pile process, the air distribution system under the pile may be one of the following:

- Disposable corrugated plastic perforated pipes that are picked up with the compost and removed by screening;
- Reusable rigid pipe pulled out of the pile before it is taken down; or
- Trenches in the floor with grating or perforated cover plate.

The spacing of pipes or trenches ranges from 1.2 to 2.4 m (4 to 8 ft). The wood chip base layer is essential to distribute air flow. In-vessel systems may use continuous plenums or gravel floors with permanent piping.

The aeration blowers may be controlled by the following means:

- Cycle timers in which the percent of time "on" is adjusted manually in response to pile temperature readings.
- Temperature feedback control in which the blowers are turned on and off automatically to maintain the pile temperature between set points.

When blowers are not running, oxygen is depleted. Murray and Thompson (1986) report significant oxygen depletion after 12 to 15 min without aeration. The "off" time should not exceed this period. Figure 5.3 shows the oxygen depletion curve reported by Murray and Thompson.

Forced aeration systems may be operated in the negative (suction) or positive (blowing) mode. The negative mode allows the core of the pile to heat up more quickly and may, therefore, be most appropriate under cold conditions or when the energy balance is marginal. In the negative mode, the exhaust gas may be more effectively captured and treated for odor removal. The positive mode drives water vapor to the pile surface and promotes drying while avoiding the accumulation of condensate in the aeration piping.

Blower motors up to 3 700 W (5 hp) may be started several times per hour. Larger motors require longer starting cycles to avoid overheating. For larger blowers, variable frequency speed controls or motorized flow control dampers may be preferable for controlling air flow.

DETENTION TIME. The time period required to stabilize the organic material typically is divided between an active composting stage and a curing stage, as shown in Figure 5.1. When the aerated static pile process was developed at the U.S. Department of Agriculture (USDA), it was found that 21 days of aerated composting followed by 30 days of unaerated curing would adequately stabilize a raw feed with wood chips as the bulking agent. This detention time criterion has been codified in a number of state regulations and incorporated into some design standards. For example, New York requires a minimum total processing time of 50 days. Most horizontal agitated bed systems are designed for 21 days of aerated composting followed by curing. However, other in-vessel systems use shorter active composting times, often 14 days.

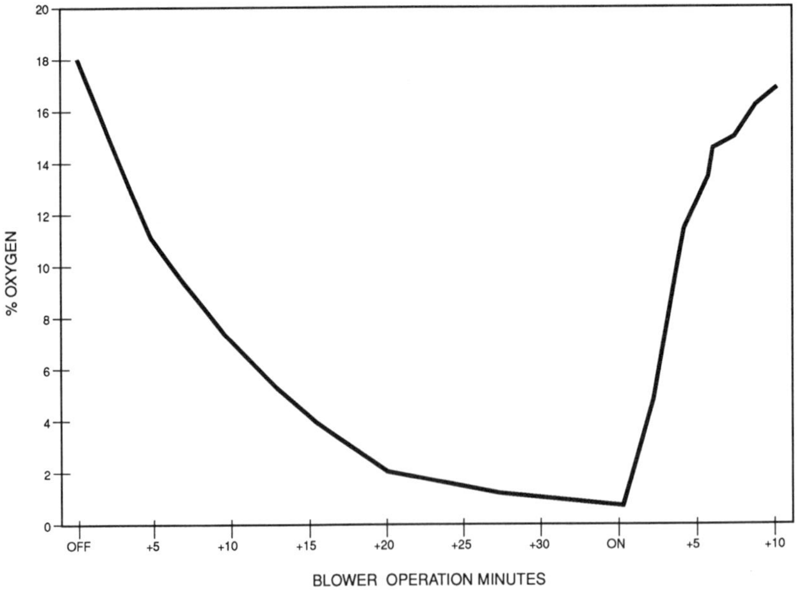

Figure 5.3   Active compost pile—oxygen depletion and regeneration (Murray and Thompson, 1986).

The detention time is affected by the bulking agent or amendment, C:N ratio, and the pH. An amendment that is not screened out may continue to decompose, requiring a prolonged curing period. An excessively high C:N ratio may have the same effect. Lime-conditioned feeds enter the process with a pH range of 9 to 11. As carbon dioxide is generated, the pH drops rapidly and there is little effect on detention time. Thermally conditioned feeds enter the process with a pH range of 5 to 6.

The composting process has no fixed end point because the organic materials continue to decompose after the compost is considered stable. One of the tests for stability is based on respiration rate measured as rate of carbon dioxide evolution. A respiration rate of 3 mg carbon dioxide per gram of organic carbon per day typically indicates that compost will be free of fecal odor and phytotoxic effects.

An alternative test measures oxygen consumption instead of carbon dioxide production. Jimenez and Garcia (1989) report that a compost taking up 0.96 mg oxygen per gram carbon per day is considered stable. This is equivalent to 1.4 mg carbon dioxide per gram carbon per day. Stability testing methodologies are moving from the development stage into routine use, but are not yet standardized throughout the composting industry.

*D*ESCRIPTION OF COMPOSTING METHODS

Although composting is a biological process occurring naturally, the degree of control imposed on a system can vary from as simple as turning a pile or windrow periodically to as involved as an enclosed or in-vessel system with mechanical agitation and odor control.

In an attempt to respond to local and regional needs, a number of composting methods have evolved. These methods offer such benefits as accelerating a naturally occurring biological process, providing for process control over variables such as moisture, carbon, nitrogen, and oxygen; containing odors and particulates; reducing land area requirements; reliably producing consistent product quality; and integrating aesthetically pleasing facilities into local and regional sites.

AERATED STATIC PILE COMPOSTING. The aerated static pile method of composting was developed in the 1970s by the USDA. The compost mixture is constructed into an approximately 2- to 2.5 m (6- to 8-ft) deep pile on top of a 0.3-m (1-ft) deep layer of wood chips. The wood chip bed contains perforated air pipe and acts as a plenum. The aeration system consists of the blower, perforated and closed pipes, and an odor control system. The entire pile is covered with an insulating blanket of wood chips or unscreened finished compost (150 to 200 mm [6 to 8 in.] in depth) to ensure that all parts of the mixture meet temperature standards and to reduce fugitive odors. Figure 5.2 presents a typical cross section and layout of an aerated static pile. In small operations, individual piles may be constructed. In large operations, a continuous pile is divided into sections representing each day's contribution.

When using the aerated static pile method the mixture remains in the pile for the active composting period, typically ranging from 21 to 28 days. Following this, the piles are broken down and the material is first screened and then moved to a curing area. It is critical that the compost solids content be at least 50 to 55% for screening. In some facilities an intensive drying step precedes screening, with a higher aeration rate than active composting. Alternatively, screening can also follow curing. Compost typically remains in a curing pile for a minimum of 30 days to further stabilize the material.

Aerated static pile composting was originally developed for outdoor sites. Most facilities constructed in the last 5 years are covered. Because of site-specific environmental concerns (such as odors and health risks), certain static pile facilities are fully enclosed.

WINDROW COMPOSTING. In windrows, the composting mix is formed into long parallel windrows with a trapezoidal or triangular cross section. The material is then periodically turned with a front-end loader or a dedicated windrow turning machine. The intent of turning is to expose the material to the air, release moisture, and loosen and fluff the material to facilitate air movement through the windrow.

Aerated windrows are constructed over air channels to protect aeration piping from the turning equipment. Air can be forced up through the windrow or pulled down through the windrow into the channel. Aerated windrows are also periodically turned. The combination of forced aeration with physical turning optimizes the composting rate and moisture release. The physical turning exposes new interfaces between particles in the mass. A typical windrow composting system is shown in Figure 5.4.

Windrow composting is performed at open outdoor sites or under cover. Windrow systems require a large amount of space compared with other composting technologies. This is primarily because of pile geometry and the required allowance between and at the end of piles for maneuvering a windrow-turning machine.

IN-VESSEL COMPOSTING. Compared with static pile and windrow composting operations, in-vessel systems promise a more stabilized and consistent product, smaller space requirements, and better containment and control of odors. A typical in-vessel composting facility is shown in Figure 5.5.

Components of In-Vessel Composting Systems. Three materials (dewatered cake, an amendment, and recycled compost) are mixed together and placed into one or several aerated reactors for composting. After composting, the product is removed from the composting reactor for curing and storage before use.

Mechanically and operationally, materials-handling systems are the dominant features of in-vessel facilities. In this regard, in-vessel facilities differ from other more common composting facilities such as aerated static pile systems. In-vessel plants are highly mechanized facilities. Designers of in-vessel facilities have taken advantage of the fact that the composting is being carried out in a reactor and that materials handling can be carried out using conveyors. There are no inherent technological requirements that demand the high degree of automation built into current in-vessel systems. It is a tradeoff of capital costs against operating costs (specifically labor).

MATERIALS-HANDLING SYSTEMS. Storage facilities for various materials are needed to provide "slack" in conveying systems so that systems operating at different flow rates or for different times can be matched. Most compost reactors are loaded for only a portion of each day, while many plants operate their dewatering facilities on a longer or different schedule. Some plants store liquid sludge in tanks and dewater it at a rate that can be accepted by the

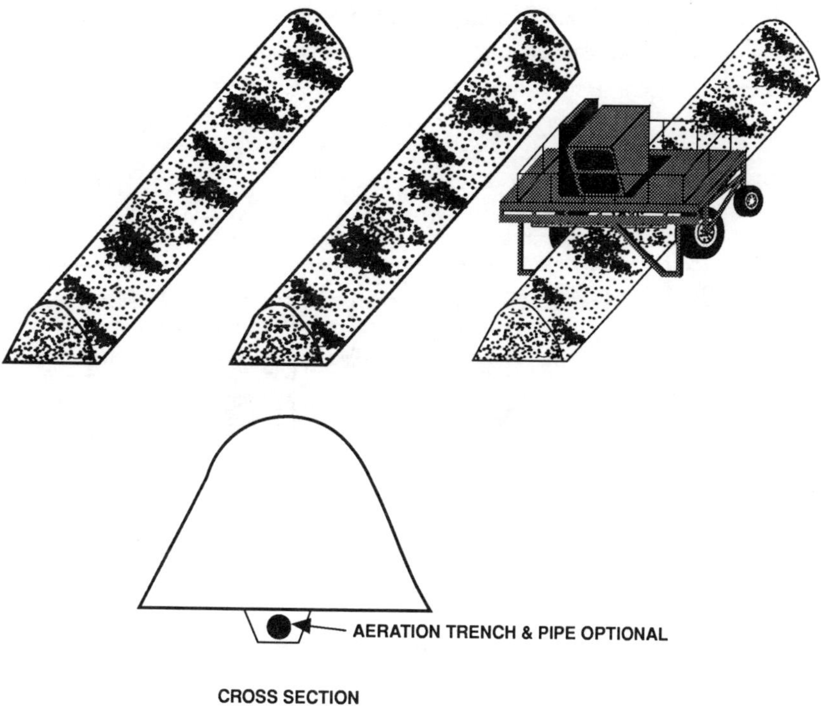

Figure 5.4   Windrow composting.

compost system. Other plants dewater during a larger portion of the day and store the dewatered cake in live-bottom bins or portable containers (if dewatering is at a site different from the composting plant).

Similarly, amendment is brought to the compost plant periodically by the truckload, but used in a continuous stream only while the reactor is loaded. In this case, storage is needed to match three different flow rates. Amendment storage facilities include covered piles, live-bottom bins, or silos similar to those used in the wood products industry.

In some plants recycle is taken from the reactor at the same time the reactor is loaded. If taken out at the exact rate needed in the mix, no storage is needed. Other plants provide recycle storage to increase the operator's flexibility and loosen the tight constraint that the reactor discharge must exactly match the mixer feedrate. Recycle typically is stored in live-bottom bins or in piles.

REACTOR SYSTEMS. The detention time in a reactor varies widely from approximately 10 to 21 days depending on system supplier recommendations, regulatory requirements, and costs. Reactor detention time should be based on desired product characteristics, especially stability, and should take into account the detention time in all process phases.

There are three general classes of reactors: vertical plug-flow reactors, horizontal plug-flow reactors, and agitated bin reactors. Vertical plug-flow reactors are vessels constructed of steel, concrete, and/or reinforced fiber glass

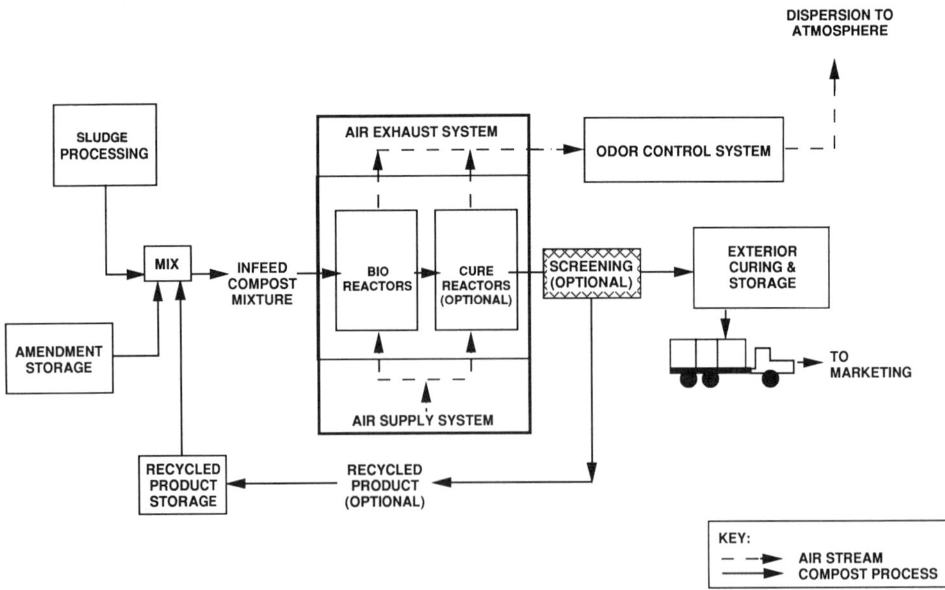

Figure 5.5   In-vessel composting facility generic process flow diagram.

panels. The mix of dewatered cake, amendment, and recycle is placed on the top of the reactor. The composting mix is aerated but not agitated or mixed. It moves as a plug to the bottom of the reactor. There it is removed by a traveling auger discharge device. Horizontal plug-flow reactors are similar to vertical reactors in that the contents are not mixed within the reactor. The compost is moved through the reactor by the action of a hydraulic ram. Agitated bin reactors differ from plug-flow reactors in that the compost does not move as an unmixed mass through the reactor. Instead, mechanical devices periodically agitate and convey forward the composting material during its stay in the reactor. Physically, the reactors are open-topped bins with air supplied with blowers from the bottom. A variety of methods are used to translocate and remove compost from the reactors. In concept these systems are similar to aerated windrows.

AERATION EQUIPMENT. Unlike materials-handling systems that use a wide variety of conveyors, relatively few types of air-moving equipment are used. The choice of equipment typically is based on the air pressure required. For air supply, positive displacement or centrifugal blowers are used where the resistance to the air flow is high. Where the backpressure is low, centrifugal fans can be used. Air supply systems are designed to provide a range of flow rates. Air flow rates are varied by changing the speed of the blower motor, throttling the inlet or operating a different number of blowers.

CURING AND STORAGE. Most in-vessel composting systems use a separate curing and storage step before product use. Sometimes curing is carried out in-vessel; however, curing typically is conducted outside on impervious surfaces. Curing facilities are sometimes aerated either with blowers in a static-pile arrangement or by turning as in a windrow facility. In locations where weather inhibits curing, covered areas are sometimes used. Detention times for these steps vary considerably at existing in-vessel facilities. However, detention times are more related to regulatory requirements and product use than to the detention in the first stage reactor. Some local regulatory agencies require 3 to 4 weeks of curing before product use. Storage times typically depend on the seasonal demand variations for the compost product.

Description of In-Vessel Systems. *VERTICAL PLUG-FLOW SYSTEMS.* In the U.S. there are three major vertical plug-flow systems in use. The primary differences among them are the configuration of the aeration systems and the discharge devices. Reactor shape, materials of construction, and ancillary equipment also vary. The depth of the composting materials is very similar, ranging from approximately 7 to 8 m (23 to 26 ft).

Reactors may be steel cylinders of 6 to 12 m (20 to 40 ft) diam capped with steel or fiber glass domes. The bottom of the reactor typically is contained in a building. A typical reactor is illustrated in Figure 5.6. Air is

introduced into the bottom of the reactor. In theory, air flows up through the
compost mix and is captured under the dome where it is drawn out by an exhaust system. The discharge device is a center-pivot traveling auger screw
located in the reactor above the aeration piping. As the auger rotates horizontally around the reactor center, it draws material to a central discharge chute.

Other reactors are enclosed in concrete rectangular structures. A typical reactor is illustrated in Figure 5.7. Air is introduced into the bottom of the reactor, which measures approximately 8 m × 5 m (26 ft × 17 ft). In theory it
flows up through the compost and is captured by a series of exhaust pipes

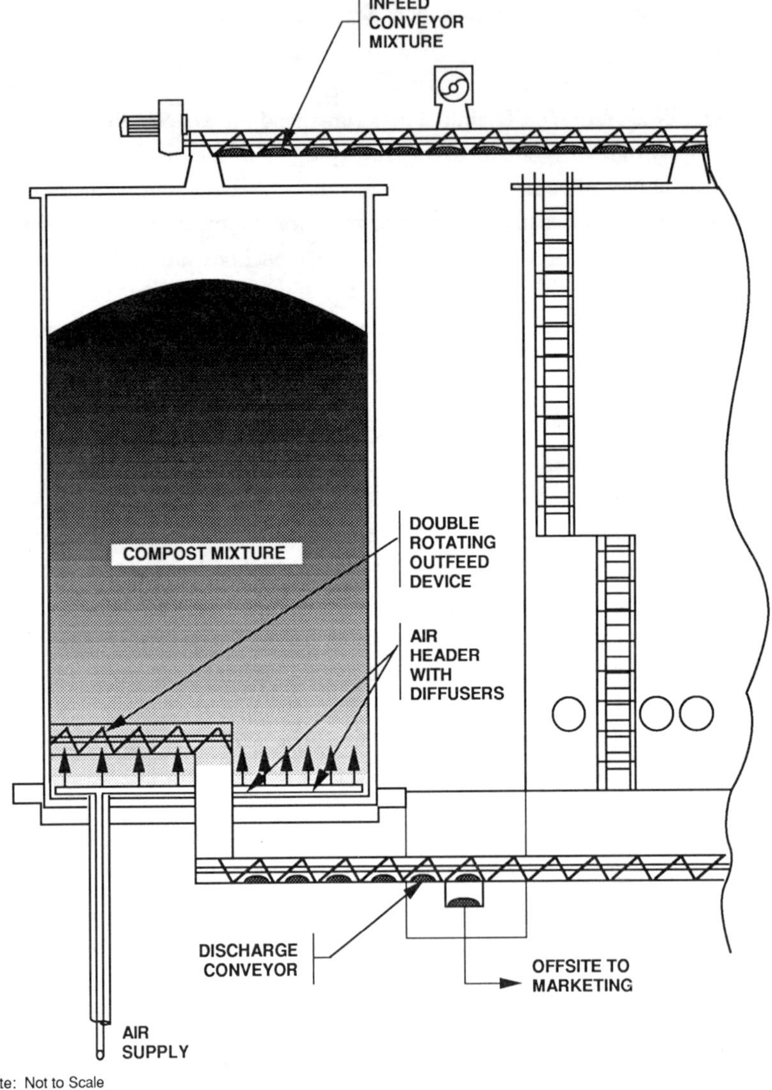

Figure 5.6   Steel cylinder vertical plug-flow reactor.

that extend approximately 1 to 1.5 m (3 to 5 ft) into the top of the compost. The rate of gas exhaust is greater than the air supply rate so that some air is pulled in the compost from above to maintain a negative pressure under the superstructure. Because of this feature, these reactors typically are covered with simple metal superstructures that are not airtight. The discharge device is an auger powered by end-mounted machinery. It passes horizontally through the compost material using a 1-m gap at the bottom of the reactor wall just above the aeration piping. To prevent escape of the reactor supply air through the gap, the room housing the discharge device must be airtight. Compost is pulled to one side and dropped onto a conveyor for removal.

Other reactors are constructed of rectangular steel frames with fiber-glass-reinforced plastic panels. The reactors, which measure 8 m^2 (26 sq ft), typically are housed in buildings for protection from the weather. A typical reactor is illustrated in Figure 5.8. Originally, air was both fed into and removed from the compost through a series of air "lances." The lances hang from headers on top of the reactor and extend to just above the discharge auger. At any one time, half of the lances are supplying air and half are exhausting air. The air path is lateral among the lances, which typically are spaced 1.5 m (5 ft) on center. Air lance systems may need to be modified so that all air lances are operated in the positive aeration mode to keep the entire reactor contents aerobic.

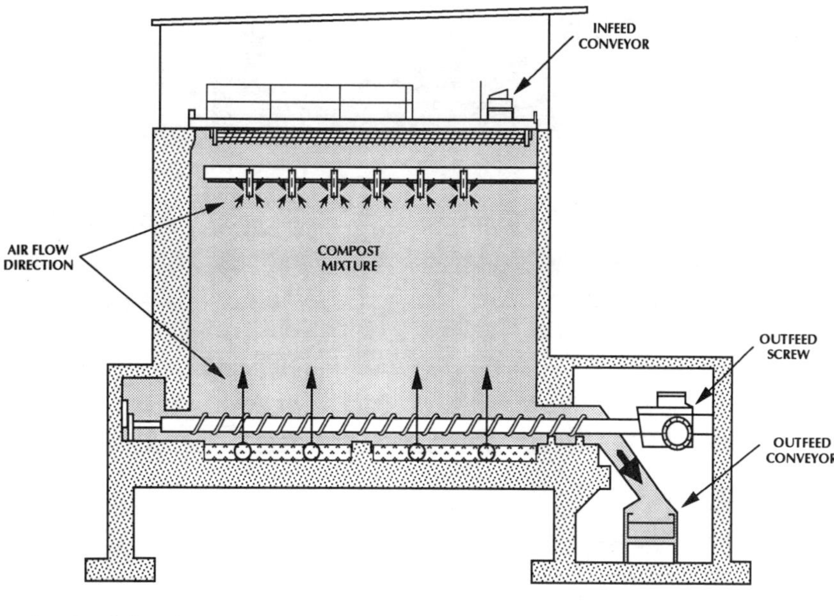

Figure 5.7   Concrete rectangular vertical plug-flow reactor.

HORIZONTAL PLUG-FLOW SYSTEMS. The second general reactor type is the horizontal plug-flow reactor, also known as the "tunnel" reactor because of its shape. The reactor typically is constructed of concrete with a pneumatically driven steel ram at one end and a drag conveyor at the other end. The modular unit is approximately 5 m (18 ft) wide, 3 m (12 ft) deep, and 20 m (65 ft) long. A typical reactor is illustrated in Figure 5.9. Except when loading, the steel ram is positioned flush against the compost. During loading, the ram is pulled back away from the compost and fresh mix is placed between it and the face of the compost bed. The ram is then closed, which compresses the mix and pushes the whole compost bed longitudinally along the reactor. At the outlet end of the reactor, finished compost sloughs off the bed as it moves forward over the edge of the collecting conveyor. Air is both supplied and exhausted through slots in the concrete floor of the reactor. The aeration system is divided into seven independent temperature zones.

AGITATED BIN REACTORS. Bin reactors differ from plug-flow reactors in the fact that the compost does not move as an unmixed mass through the reactor. Instead, mechanical devices agitate the composting materials periodically during its stay in the reactor. Physically, the reactors are open-topped bins

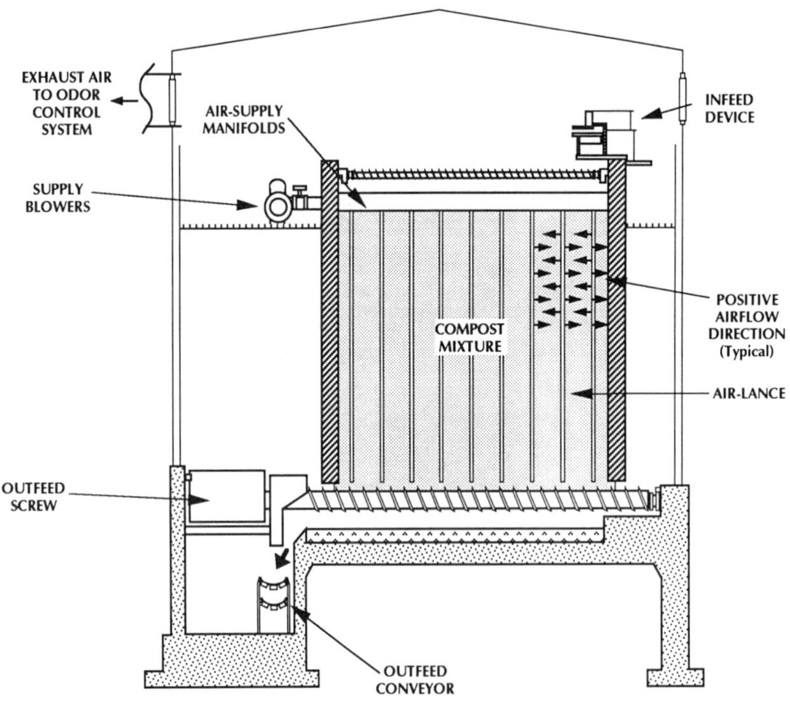

Figure 5.8   Steel rectangular vertical plug-flow reactor.

with air supplied from the bottom. The depth of the compost beds is 2 to 3 m
(6 to 10 ft). Bin reactors can be either rectangular or circular.

One typical rectangular reactor is illustrated in Figure 5.10. The reactors
are enclosed in a building. Air is exhausted from the compost bed into the
building atmosphere or it is drawn downward through the bin and collected
for odor control. The aeration system is divided into a number of independent
aeration zones along the bottom of the reactor. The compost is both agitated
and removed from the reactor by a mobile extraction/conveyor device that
spans the compost reactor and rides on rails mounted on the reactor walls.
This device operates semiautomatically, controlled by an operator in an enclosed cab on the device. It is equipped with rotating toothed drums that agitate the composting material and pull it onto an inclined table conveyor,
which lifts it out of the reactor. In the agitation mode, the extraction device
drops the compost off the table conveyor, letting it fall back into the reactor.
Agitation typically is done weekly. In the discharge mode, the extraction device discharges the compost onto a mobile transfer conveyor that moves behind it. The reactor is loaded by a traveling conveyor similar in construction
to the transfer conveyor.

An alternative horizontal agitated bed reactor is illustrated in Figure 5.11.
These reactors are rectangular, aerated from the bottom with independently
programmable aeration zones and enclosed in a building. Fresh mix is loaded
into the front end by a loader. The agitation device is completely automatic
and operates only in the agitation mode. The agitation device typically makes
one pass through the reactor each day. The composting material is dug out
and redeposited approximately 4 m (12 ft) behind the machine. Eventually,
the composting material is moved through the entire length of the reactor.

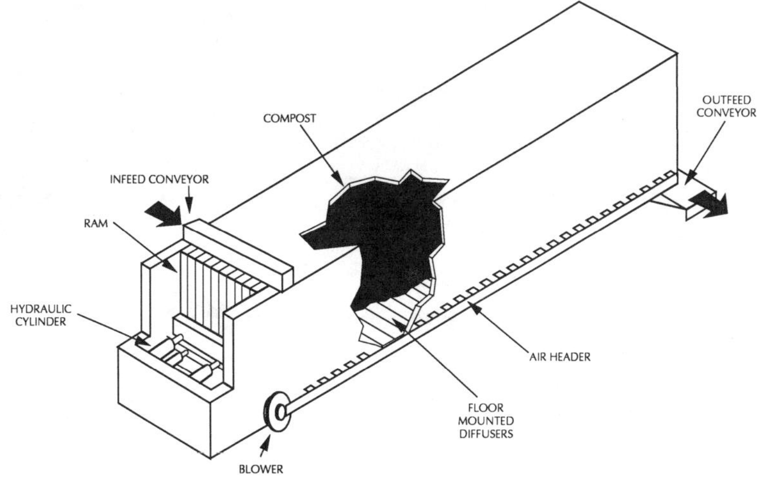

Figure 5.9   Horizontal plug-flow reactor.

Another type of reactor is a circular vessel approximately 3 m (10 ft) deep, typically covered with an aluminum geodesic dome. Its diameter can vary up to approximately 36 m (120 ft). Air is supplied from the bottom of the reactor and exhausts into the space between the top of the compost bed and the underside of the dome, where it is vented away. Ventilation in the negative aeration mode is also feasible with this reactor. The aeration system is divided into five independent annular zones. A typical reactor is illustrated in Figure 5.12. Agitation is provided by vertical augers mounted on half of a rotating bridge that spans the compost bed and rides on a rail mounted on the reactor wall. The augers extend to about 150 mm (6 in.) above the aeration piping. The augers are mounted at a slight inward angle so that as they mix the compost they also move it from the perimeter to the center of the reactor. At the center is a circular discharge chute that accepts finished compost from the innermost auger. Fresh mix is loaded on the perimeter of the reactor by a conveyor on the opposite side of the bridge from the augers. Agitation of the compost bed is not continuous. It occurs only during loading operations.

COMPARISON OF COMPOSTING METHODS. Composting has been growing in popularity in the U.S. According to a 1993 *BioCycle* survey, there were a total of 207 composting projects in operation and under construction, as shown in Table 5.5. The survey shows that approximately 50% of the operational facilities are static pile, with the remainder split between windrow

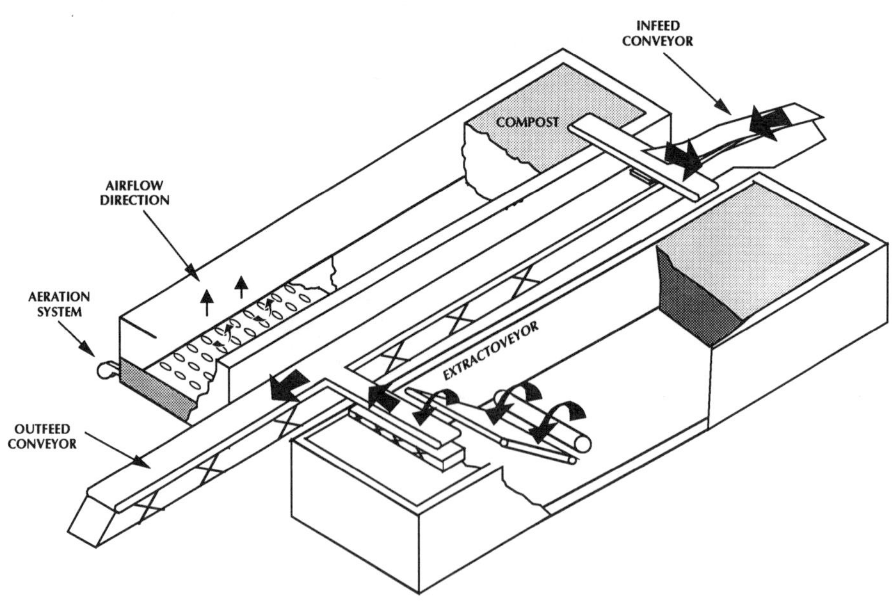

Figure 5.10   Horizontal mixed reactor (plan view).

and in-vessel facilities. Several reasons have been given to account for the growth of composting, including the following:

- Decreasing availability of landfills for solids disposal,
- Tipping fees for landfills and other solids management alternatives have reached a level that makes the economics of composting feasible, and
- Greater emphasis being placed on beneficial reuse at federal, state, and local levels.

A representative list of operating facilities is shown in Table 5.6. This list of facilities represents several types of composting systems in use throughout the U.S. The list includes some of the more well-established facilities with very successful compost marketing programs.

Table 5.7 presents a comparison of the advantages and disadvantages of five composting technologies based on consideration of physical facilities, processing aspects, and operation and maintenance. No one of the technologies is universally appropriate for every situation. Climate, siting consideration, operational concerns, sensitivity to odors, and other factors will influence the choice of systems. The factors considered important in deciding among the composting technologies follow.

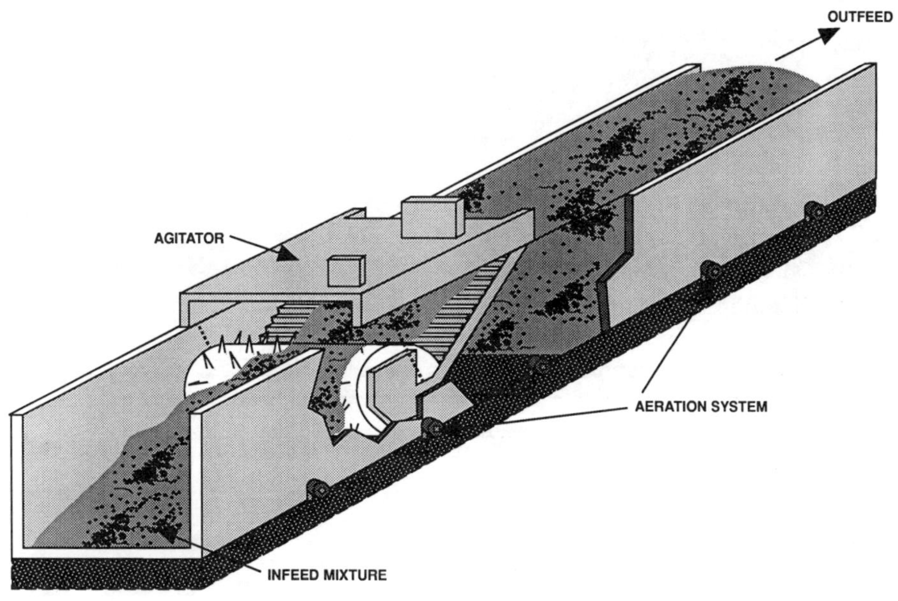

Figure 5.11 Horizontal agitated bed reactor.

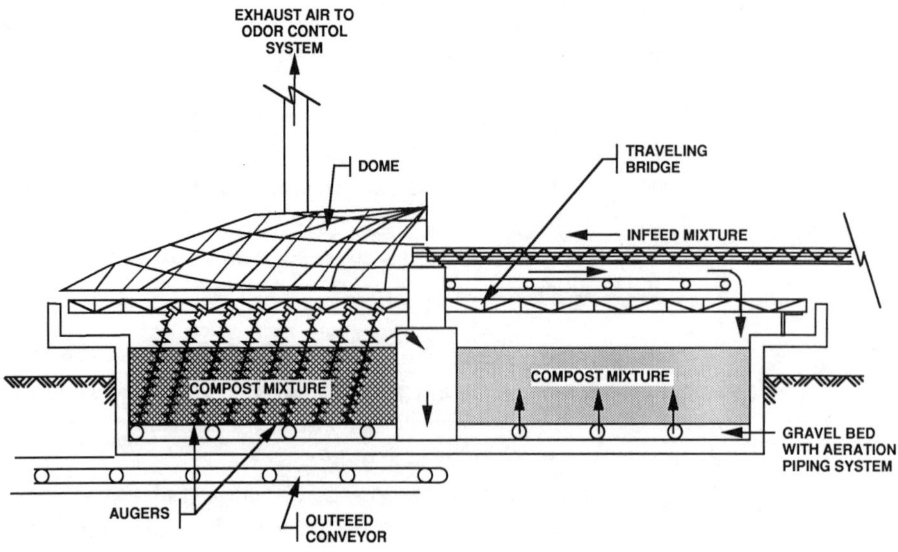

Figure 5.12   Horizontal mixed reactor (side view).

- Physical facilities
  — Area requirements,
  — Materials handling system complexity,
  — Aeration equipment, and
  — Degree of enclosure.

- Process considerations
  — Uniform aeration,
  — Adaptability to different bulking agents,
  — Adaptability to changes in volume of feed solids,
  — Odor emissions/odor control, and
  — Plug-flow versus mixed technologies.

Table 5.5   Summary of composting facilities (*BioCycle*, 1993).

Type	Operational	Construction
Aerated static pile	92	8
Static pile	4	0
Aerated windrow	14	0
Windrow	38	3
In-vessel	36	10
Unspecified	2	0
Total	186	21

- Operation and maintenance
  - — Labor requirements,
  - — Operator exposure,
  - — Dust generation, and
  - — Degree of maintenance.

ODOR CONTROL

Control of odors is perhaps the greatest challenge faced by the composting industry. Most suspensions of plant operations and most conflicts over plant siting have been caused by odors or concern over future odor effects. The composting process is inherently odorous because the volatile products of decomposition are removed by a flow of air. There is increased emphasis in plant design and operations on enclosure, capture, and treatment of exhaust air, and improved process control to reduce odors at the source.

ODOR SOURCES IN COMPOSTING. Every stage in the composting process is a potential source of odors. Odor sources may be divided into the following three categories:

Table 5.6 Representative list of composting facilities in the U.S.

Facility name and/or location	Type	Capacity, dry ton/d[a]
East Bay MUD, Oakland, California	Aerated static pile	14
Denver Metro, Denver, Colorado	Aerated windrow	10
Longmont, Colorado	Aerated static pile	4
Hartford, Connecticut	In-vessel	50
Palm Beach County, Florida	In-vessel	3
Washington Suburban Sanitary Commission, Montgomery County, Maryland	Aerated static pile	40
Springfield, Massachusetts	In-vessel	40
Dover, New Hampshire	Aerated static pile	5
Lockport, New York	In-vessel	10
Hamilton, Ohio	In-vessel	15
Hampton Roads Sanitary District, Virginia	Aerated static pile	12
Austin, Texas	Windrow	10
Albuquerque, New Mexico	Aerated windrow	15

[a] ton/d $\times$ 0.907 2 = Mg/d.

Table 5.7   Key advantages and disadvantages of composting systems.

Composting technology	Advantages	Disadvantages
Aerated static pile	• Adaptability to various bulking agents • Flexibility to handle changing feed conditions and peak loads (volume not fixed) • Relatively simple mechanical equipment	• Relatively labor intensive • Relatively large area required • Operators exposed to composting piles • Potentially dusty working environment
Windrow	• Adaptability to various bulking agents • Flexibility to handle changing feed conditions and peak loads (volume not fixed) • Relatively simple mechanical equipment • Requires no fixed mechanical equipment	• Very large area required • Relatively labor intensive • Operators exposed to composting piles • Dusty working conditions
Vertical plug flow	• Completely enclosed reactors in some systems improve ability to control odors • Relatively smaller area required • Operators not exposed to composting material	• Single outfeed device per reactor (large reactors), potential bottleneck • Potential inability to maintain unform aerobic conditions throughout reactor • Relatively maintenance intensive • Limited flexibility to handle changing conditons • Materials-handling system may limit choice of bulking agents
Horizontal plug flow (tunnel)	• Completely enclosed reactors improve ability to control odors • Relatively smaller area required (compost mix compacted) • Operators not exposed to composting material	• Fixed volume reactors (no flexibility) • Limited ability to handle changing conditions • Relatively maintenance intensive • Materials-handling system may limit choice of bulking agents
Agitated bin	• Mixing enhances aeration and uniformity of compost mixtures • Ability to mix compost (advantage in handling some bulking agents) • Adaptabilty to various bulking agents	• Fixed volume reactors (no flexibility) • Relatively large area required • Potentially dusty working environment • Operators exposed to composting piles • Relatively maintenance intensive

- Active sources—sources of odor that exist when material is being actively handled such as mixing, screening, and dewatering. Odors from these sources occur during working hours.
- Continuous sources—sources originating in the aeration and storage of materials. These may be point sources, such as blower exhaust, or pile and windrow surface emissions. Odors may occur 24 hours per day.
- Housekeeping sources—every activity or stage has associated opportunities for material spills, unclean equipment, and leachate on ground surfaces. These can be major odor sources that persist after daily activity has stopped and are, therefore, continuous sources.

Table 5.8 illustrates typical odor sources by the three categories. Table 5.9 lists the most common odor compounds, published odor detection threshold concentrations, and pathways of formation. Detection threshold concentrations are variable, and published ranges may span several orders of magnitude.

ODOR MEASUREMENT. Concentrations of individual compounds can be measured by standard analytical methods. A simple apparatus consisting of a syringe and colorimetric adsorption tube can be used in the field. Tubes are available for a number of the compounds listed in Table 5.9. More accurate and complete analysis can be performed by collecting samples and analyzing in the laboratory by gas chromatography. Gas samples may be collected in bags, stainless steel vacuum canisters, or tubes filled with adsorbant.

A composting odor typically is a mixture of compounds that cannot be quantified as a sum of individual constituents. Odors can be measured directly only by the human nose (sensory analysis). However, odor samples can be captured in Tedlar® bags for sensory analysis at another location. Several methodologies have been developed for quantifying concentration of odor using a panel of human subjects.

Odor Concentration. Odor concentration as measured in dilutions to threshold, or ED_{50}, is defined as the number of dilutions of an odorous sample required to render the sample undetectable to one half of the panelists. This parameter is measured by subjecting an odor panel of approximately 10 members to various dilutions of the odor, starting with the most dilute. Each panelist is presented with three sniffing tubes. Two are connected to odor-free air and one to the diluted sample. Each panelist is required to identify the tube with the sample. The dilution at which 50% of the panelists can correctly identify the sample tube is the ED_{50} (ASTM Standard 679).

Odor Intensity. Odor intensity is measured by comparing the odor to a "standard odor" comprised of a known concentration of *n*-butanol in air. The

intensity is assigned a rating of one to eight, eight being the most intense, or
is assigned a butanol equivalent in parts per million by volume (ppmv) in the
air. The procedure is standardized in ASTM Standard 544.

By measuring the intensity of various dilutions of the odorous sample, the
relationship between intensity and dilution can be determined. The slope of a
log–log plot of intensity versus dilution is a relative measure of the pervasive-
ness of the odor. A steep slope (high value) indicates an odor whose intensity

Table 5.8   Typical odor sources in composting operations.

Odor source	Category*
Dewatered sludge transport and storage	
Open trucks en route	A
Open trucks parked on site	A
Dumping operations	C
Untreated ventilation from storage facilities	C
Open conveyors	A
Spillage from trucks	H
Spillage around storage facilities	H
Residue on empty trucks	H
Puddles from truck washing	H
Tire tracking of spilled sludge	H
Mixing	
Surface emissions from mixing by front-end loader or batch mixer	A
Untreated ventilation from pug mills	A
Mix left on paved surface after day's activities	H
Residue on equipment	H
Pile building	
Surface emission from materials-handling activities	A
Spillage of mix left on paved surface after day's activities	H
Residue on equipment	H
Sludge balls from poor mixing	H
Surface emissions from pile before placement of blanket layer	A
Composting	
Surface emissions from active piles	C
Leachate puddles at base of piles	H
Aeration	
Blowers exhaust	C
Leakage of condensate from aeration piping	H
Leakage of exhaust from piping and blower housing	H

*A = active sources; C = continuous sources; and H = housekeeping sources.

Table 5.9 Odor compounds and sources (Verschueren, 1983, and WPCF, 1979).

Class compounds	Odor threshold, ppm	Likely source at treatment plant	Pathway of formation/release
Inorganic sulfur			
Hydrogen sulfide	0.000 47	Septic wastewater or sludge	Anaerobic reduction of sulfate to sulfide or anaerobic breakdown of amino acids
Organic sulfur			
Mercaptans			
Ethyl mercaptan	0.000 19	Sludge or wastewater subjected to anaerobic conditions	Anaerobic and aerobic breakdown of amino acids
tert-Butyl mercaptan	0.000 08		
Allyl mercaptan	0.000 05		
Organic sulfides			
Dimethyl sulfide	0.001	Composting	Aerobic oxidation of mercaptans
Dimethyl disulfide	0.002		
Inorganic nitrogen			
Ammonia	0.037	Composting; processing of anaerobically digested biosolids	Anaerobic decomposition of organic nitrogen; volatilization at high pH, temperature
Organic nitrogen			
Methylamine	0.021	Solids processing	Anaerobic decomposition of acids
Ethylamine	0.83		
Dimethylamine	0.047		
Fatty acids			
Acetic acid	0.001–1.0	Sludge subjected to anaerobic conditions	Anaerobic decomposition
Propionic acid	0.005–0.05		
Gutyric acid	0.000 01–0.01		
Aromatics			
Acetone	0.05–400	Preliminary and primary wastewater treatment processes, solids processing, and composting	Present in wastewater contribution from industry; breakdown of lignins
Methylethyl ketone	1–12	Composting, wood-based bulking agents	
Terpenes	Varies		Present in wood products, such as wood chips, sawdust

is reduced rapidly on dilution (not pervasive), while flat slope (low value) indicates an odor that retains a high intensity on dilution (pervasive).

Table 5.9 indicates that ammonia has a higher threshold of detectability than organic sulfur compounds. Ammonia may be the dominant odor at the plant, masking the organic sulfur odors. Downwind, the ammonia concentration is below detectability and the organic sulfur odors will predominate at extremely low concentrations.

ATMOSPHERIC DISPERSION AND FACILITY SITING. The dispersion and dilution of odors is dependent on the following factors:

- Distance to receptor,
- Topography,
- Atmospheric stability and wind speed, and
- Odor emission rate and discharge conditions.

An odor plume spreads out horizontally and vertically with distance; some dilution may be provided within the first 300 m (1 000 ft). Under stable atmospheric conditions, dilution from ground-level discharges may not be sufficient to prevent odor effects. A composting facility with a large buffer distance can be designed with less odor control and is less likely to generate complaints. Many WWTPs are located in river valleys. Valleys are poor locations for odor sources because atmospheric mixing is limited. Odors tend to move along the valley. There are practical advantages to locating the composting operation at the WWTP. These advantages should be weighed against the need for increased odor control and probability of odor complaints.

Stability is an inverse representation of the degree for atmospheric mixing. Stability Class A indicates rapid vertical and horizontal mixing. The poorest mixing condition is Stability Class F. Odors are frequently most noticeable early on cold mornings, when cold air at ground level traps odors and cannot rise. As the sun warms the air, there is vertical mixing and odors disperse. As wind speed picks up, odors move in "puffs." At a particular location, the odor may suddenly become intense for a few minutes and then subside. People are sensitive to odor changes, so puffs may elicit complaints.

Odor emission rate is the product of exhaust air flow rate and odor concentration or intensity. Downwind odor intensity is proportional to odor emission rate. It is therefore desirable to reduce the amount of air that must be discharged. A tall stack or high stack velocity will improve dispersion.

A number of models are available to predict atmospheric dispersion. They range in complexity from simple nomographs to models that account for topography and statistically predict the frequency and intensity of odor puffs. At the point of generation, odor concentration is measured as dilutions to threshold. Models predict the degree of dilution. It is then possible to design the control system so that odors are below threshold at the nearest receptor or property line under most atmospheric conditions.

ODOR CONTROL THROUGH OPERATIONS. Table 5.10 lists typical sources of excess odor in composting operations. The table shows that many odor sources are results of poor housekeeping. Operational improvements may eliminate significant odor emissions at low cost. Process control improvements may also eliminate significant odor emissions. Table 5.10 identifies typical process control deficiencies that may cause odor.

CONTAINMENT AND TREATMENT. Operational and process improvements may provide adequate reduction in odor effects. However, the trend in new facilities is toward containment and treatment of exhaust. The following odor treatment technologies typically are used in composting facilities.

Table 5.10   Process-related causes of excessive odors.

Mixing
- A mix that is too wet or dense will not have adequate porosity.
- A mix with a low C:N ratio may generate excessive ammonia (remedied through the quantity and quality of the bulking agent).
- A nonhomogeneous mix will result in nonuniform air flow.

Aeration
- Inadequate aeration or long "off" cycles on blower controls will result in anaerobic conditions.
- Inadequate aeration will result in high temperatures that may increase odor generation.

Pile construction (static pile)
- If piles are too deep, air may become saturated and oxygen deficient before penetrating the full depth.
- The blanket layer acts as an odor filter and must be placed on all active piles.

Screening
- If recovered bulking agent contains too much compost, the compost may reduce porosity and C:N ratio. This is remedied by ensuring that the compost is dry enough to screen and that the screen is not overloaded.
- If the compost contains too much fine bulking agent, it may result in high temperatures and odor generation during curing. This may be remedied by a fine screen or prescreening bulking agent.

Curing
- If compost is still highly unstable or curing piles are too deep, anaerobic conditions and high temperatures may occur. This may be remedied by upstream changes, smaller piles, or aeration of curing piles.

Biofilters or Earth Filters. Beds of blended media consisting of materials
such as mature compost, soil peat, and bark, are effective in removing odors,
particularly reduced sulfur compounds. Odor compounds are adsorbed and
oxidized by microorganisms. The loading rate typically is 0.025 $m^3/m^3 \cdot s$
(5 cfm/sq ft) or less; therefore, large land areas are required for high flow
rates. Biofilters must be kept moist and may be sensitive to high ammonia
concentrations. They are relatively inexpensive, easy to operate, and highly
effective if constructed and maintained correctly. Biofilters typically discharge at ground level, providing less atmospheric dispersion than stacks.
When sensitive receptors are close to the facility, dispersion can be enhanced
by enclosing the biofilter and discharging through a stack.

Wet Scrubbers. Multistage atmozing or packed tower scrubbers are used successfully for odor control at composting facilities. The first stage typically
uses water and acid to remove ammonia. A surfactant or detergent may be
used in the first stage to improve capture of nonpolar compounds, such as ter-
penes. The second stage uses sodium hypochlorite to oxidize organic compounds. The most effective scrubbers have a third stage for dechlorination
and polishing, typically using hydrogen peroxide.

OPERATIONS

The general principles of compost facility operations are derived from basic
theory, design criteria, odor control, and product quality requirements.

DEWATERING. A small increase in cake solids content allows a significant
reduction in the ratio of bulking agent to cake and corresponding decrease in
material handling requirements.

BULKING AGENT. The solids content of the bulking agent also significantly affects the ratio of bulking agent to cake. Less bulking agent is required
for higher cake solids.

MIXING. The mixture should have a minimum solids content of approximately 40% for static pile systems. Agitated bed systems may work with
slightly lower solids content. Cake must have high surface-to-volume ratio
and close contact with bulking agent. Dewatered cake of 20 to 25% solids
should form a thin coating on the bulking agent particles. Dry, crumbly cake
should be broken into fine particles and mixed well with bulking agent. Balls
and clumps will not compost and will contribute to odors. The mix must be
highly uniform with respect to porosity or density. Uniformity is difficult to
achieve with a front-end loader. If a loader is used, the mix should be made in

batches of only a few bucket loads. Uniformity is most critical in static pile systems and less so in windrow and agitated bed systems as they are periodically remixed.

TEMPERATURE CONTROL AND AERATION. Temperature should be measured daily at several representative points in the volume associated with each blower or aeration zone. Unless the facility has temperature feedback control, the "on" time of the blower cycle should be adjusted several times per week in response to temperature trends. The "off" time of the blower cycle should not exceed 12 minutes to avoid oxygen depletion.

The temperature should be kept between 55 and 65°C until pathogen control requirements are met. If the temperature cannot be kept at or less than 65°C, the aeration capacity is inadequate. After pathogen control requirements are met, aeration can be increased to maintain a temperature of 50 to 55°C, which is the range of most rapid stabilization and drying. Uniform air distribution is critical.

In the static pile process, the insulating layer is critical to ensure pathogen inactivation near the pile surface and to capture odors. In the static pile process, negative (suction) aeration promotes initial temperature rise. Positive aeration promotes cooling and drying.

Exhaust from negative aeration should be treated for odor removal. Condensate may accumulate in the aeration system and block air flow.

SCREENING. The compost solids content must be at least 50 to 55% to prevent blinding of the screen. A solids content of more than 65% may cause dust and will inhibit curing. The screen must not be overloaded, to avoid carryover of the fine fraction.

RECORDKEEPING. In most composting processes operational data are manually generated. The recordkeeping system should be designed to meet the following objectives without placing an excessive burden on the operators:

- Temperature data to document pathogen reduction;
- Analytical data required by regulatory agencies and end users; and
- Input and output to track materials flow and bulking agent costs, dewatered cake and bulking agent solids content, volatile solids and nitrogen mix ratios, and mix densities.

In-vessel facilities and many static pile facilities use thermocouple probes to monitor and record pile temperatures and to control the blower operations. Software systems are available that provide graphic displays of mixing, screening, conveyors, and handling equipment status. These systems generate records of temperature and quantities of material processed each day.

SAFETY. Operational procedures must emphasize safety and worker health. Rotating machinery is potentially hazardous. Areas where front-end loaders and other moving equipment are operating are also hazardous. Blowers may present electrical hazards if they are wired through cords or if there is standing water. Wet and slippery conditions should be prevented and remedied. Workers should wear dust masks when digging into piles or operating screening equipment. Effective housekeeping procedures are essential to worker safety, odor control, and public acceptance of the composting facility.

COMPOST MARKETING AND UTILIZATION

Compost is a soil conditioner with the following characteristics:

- Increases water holding capacity for sandy soils;
- Increases aeration and drainage for clay soil;
- Provides slow release nutrients of nitrogen (1 to 2%), phosphorus (2%), and potassium (0 to 1%);
- Provides essential plant micronutrients; and
- Increases nutrient holding capacity of soil.

In the composting process, much of the nitrogen is converted to ammonia and volatilized. Compost contains less nitrogen than the input feed cake or a heat-dried product.

QUALITY REQUIREMENTS. Compost use is regulated by federal or state requirements and market requirements. Federal regulations specify high quality and ceiling limits for inorganics and minimum quality requirements for pathogen and vector attraction reduction.

Several state regulations establish classes of compost based on inorganic concentrations. The compost of the highest (best) class may be distributed to the public. In some states it must be labeled to indicate whether it is suitable for crops grown for direct human consumption. Lower classes of compost have more use restrictions and may require site-based permitting for use.

Marketing requirements may vary according to the intended end use and the demand for compost. Some general market requirements are listed below:

- Stability or maturity
  — Minimal odors,
  — Minimal phytotoxic effects because of incomplete decomposition, and
  — Minimal nitrogen depletion because of unstabilized carbon.

- Moisture content
  - Excessive moisture adds to transportation cost and may generate odor,
  - Insufficient moisture will cause dust, and
  - Ideal moisture range is between 35 and 45%.

- Texture
  - Coarse compost may be most suitable for land reclamation, and
  - Fine compost may be required for topsoil blending and topdressing of turf.

- Salinity or conductivity
  - Excessive salinity will inhibit sprouting of seeds, and
  - Salinity may be caused by ferric chloride conditioning.

TYPICAL USES FOR COMPOST. Table 5.11 lists typical uses for compost and markets associated with each use. Uses in which compost serves as a seed germination medium will be particularly sensitive to phytotoxicity and salinity. Topdressing applications will be sensitive to particle size. For landfill cover and land reclamation, coarser compost may be preferred, because it will be more resistant to erosion. Retail or homeowner applications will be particularly sensitive to texture and residual odor. Many applications are sensitive to nitrogen depletion. Frequently, fertilizer is applied with the compost to ensure adequate nitrogen.

APPLICATION RATES. Annual application rates typically are limited by the nitrogen uptake of the plants. The nitrogen in compost is released gradually over a period of years, so it is possible to apply amounts in excess of annual plant uptake by applying at an interval of more than 1 year.

Established turf may be topdressed with a layer of mature, finely screened compost at a rate of 6 to 10 mm per application, applying every 2 to 3 years. This is equivalent to 4 to 6 Mg/ha/application (2 to 3 ton/ac/application). Up to 100 mm (4 in.) of compost may be incorporated to the soil for turf establishment or land reclamation. This heavy application is not repeated.

Cumulative application rates are limited by allowable heavy metal concentrations in soil. Metals are immobilized by the organic material in the compost, but gradually become more available as the organic material breaks down in the soil.

MARKET DEVELOPMENT. The 1993 *BioCycle* survey of composting facilities reports that most producers of compost are successful in finding markets. Approximately 80% of the number of facilities surveyed reported that compost was sold in bulk form. Typical selling price was \$4 to \$8/m^3 (\$5 to \$10/cu yd). Prices per 1- or 2-lb bag were typically \$1 to \$3.

Municipal producers of compost may use the following market strategies, individually or in combination:

- Use of compost within the public sector,
- Direct marketing to users, or
- Employing a compost broker.

Often, public sector uses are developed first. Direct marketing has the advantage of maximizing revenue and visible public benefit. Brokers offer lower revenue, but they provide market security and assume much of the administrative burden of compost marketing. According to the *BioCycle* survey, approximately 25% of the municipalities reported contracts with private brokers.

Typically, compost market studies are performed during the facility planning stage. Users within the geographical area are contacted and potential rates of use are estimated. These potential rates are then scaled down by assigning conservative rates of market penetration. Aggregate demand is then estimated for each user category. Through these contacts, specific user concerns and requirements are compiled. Market studies also provide information on user constraints and type of products desired.

In many locations, the rate of compost production is expected to increase dramatically. Emerging sources of compost include municipal yard waste

Table 5.11   Typical uses and markets for compost.

Uses	Markets
Potting and horticulture mixes	Greenhouses, nurseries, retail distribution
Soil replacement	Field nurseries, sod farms
Blending to produce topsoil	Topsoil blenders, landscape material suppliers
Turf establishment	Landscape and site contractors, public sector,* retail distribution
Top-dressing of turf	Golf courses, institutions, public sector, retail distribution
Amendment of sandy or clay soils	Agriculture, also soil replacement, topsoil, turf establishment, and land reclamation markets
Land reclamation	Landfill operators, mine and gravel pit operators, landscape and site contractors
Landfill cover	Landfill operators
Private gardens	General public

* Public sector may include municipal, county, state, and federal agencies (such as parks departments, highway and public works departments, and airports).

composting operations, solid waste composting, and private producers. An effective compost market study must consider the effect of future competition from other sources. After a composting facility is in operation, maintenance and development of markets becomes an ongoing administrative responsibility. Experience at existing facilities suggests that markets are easily developed and maintained if the supply of compost is reliable and the quality is consistent.

In general, revenue from compost sales will not cover production costs, but will cover costs associated with market development. The amounts charged to commercial users and brokers by municipal producers typically have reflected market conditions rather than production costs. Where municipal compost producers have given away compost free of charge, distribution typically is limited to municipal agencies and individual users.

*R*EFERENCES

BioCycle Survey (1993).

Chang, Y. (1967) The Fungi of Wheat Straw Compost: Part II—Biochemical and Physiological Studies. *Trans. Br. Mycol. Soc.,* **50,** 667.

Goldstein, N., and Stenteville, R. (1993) Biosolids Compost Makes Healthy Progress. *BioCycle.*

Golueke, C.G. (1977) *Biological Reclamation of Solid Waste.* Rodale Press, Emmaus, Pa.

Haug, R.T. (1980) *Compost Engineering.* Ann Arbor Science Publishers, Ann Arbor, Mich.

Higgins, A.J., *et al.* (1982) Airflow Resistance in Sewage Sludge Composting Systems. *Trans. Am. Soc. Agric. Eng.,* 1010.

Iacoboni, M.D., *et al.* (1980) Deep Windrow Composting of Dewatered Sewage Sludge. *Proc. Natl. Conf. Munic. Ind. Sludge Composting Hazard. Mater. Control Res. Inst.,* Silver Spring, Md.

Jimenez, E.I., and Garcia, V.P. (1989) Evaluation of City Refuse Compost Maturity: A Review. *Biol. Wastes,* **27,** 115.

Kayhanian, M., and Tchobanoglous, G. (1992) Computation and Importance of Carbon to Nitrogen (C/N) Ratios for Various Organic Fractions of Municipal Solid Waste. *BioCycle.*

Knoll, K.H. (1964) Information Bulletin No. 13-20. Int. Res. Group on Refuse Disposal, U.S. Public Health Serv., Rockville, Md.

LeBrun, T., Los Angeles County Sanitation District (1979). Memorandum to the LA/OMA Project on Status of *Aspergillus* Monitoring.

Millner, P.D., *et al.* (1977) Occurrence of *Aspergillus fumigatus* During Composting of Sewage Sludge. *Appl. Environ. Microbiol.,* **34,** 6.

Morgan, M.T., and MacDonald, F.W. (1969) Tests Show MB Tuberculosis Doesn't Survive Composting. *J. Environ. Health,* **32,** 101.

Murray, C.M., and Thompson, J.L. (1986) Strategies for Aerated Pile Systems. *BioCycle,* **6.**

Poincelot, R.P. (1975) *The Biochemistry and Methodology of Composting.* CT Agr. Exp., N.H.

Shell, G.L., and Boyd, J.L. (1969) Composting Dewatered Sewage Sludge. U.S. Dep. of Health, Education, and Welfare, SW-12c.

Stutzenberger, F.J. (1971) Cellulase Production by *Thermomonospora curvata* Isolated From Municipal Solid Waste Compost. *Appl. Microbiol.,* **22,** 2, 147.

Verschueren, K. (1983) *Handbook of Environmental Data on Organic Chemicals.*

Waksman, S.A., and Cordon, T.C. (1939) Thermophilic Decomposition of Plant Residues in Composts by Pure and Mixed Cultures of Microorganisms. *Soil Sci.,* **47,** 217.

Water Pollution Control Federation (1979) *Odor Control for Wastewater Facilities.* Manual of Practice No. 22, Washington, D.C.

Water Pollution Control Federation (1984) *Sludge Disinfection: A Review of the Literature.* Washington, D.C.

Wiley, J.S., and Westerberg, S.C. (1969) Survival of Human Pathogens in Composted Sewage. *Appl. Microbiol.,* **18,** 944.

Yanko, W. (1987) *Occurrence of Pathogen in Distribution and Marketing of Municipal Sludge.* EPA-600/1-87-014, U.S. EPA.

Chapter 6
Alkaline Stabilization

The addition of alkaline chemicals is a reliable method of stabilization that has been practiced at wastewater treatment plants (WWTPs) since the 1890s. Quicklime (CaO) and hydrated lime [Ca(OH)$_2$] are the traditionally used alkaline additives.

In recent years, a number of advanced alkaline stabilization technologies have emerged that use either chemical additives in addition to or instead of lime, or involve special equipment or processing steps. These processes all claim advantages over traditional lime stabilization, including enhanced pathogen control and a more publicly acceptable end-use product. The end-use product produced by the advanced alkaline stabilization technologies is sometimes referred to as "artificial soil," because it has been successfully used as a soil substitute. This chapter will discuss traditional and advanced alkaline stabilization technologies.

Lime is the most widely used and one of the lowest cost alkaline materials available in the wastewater industry. Lime has been used for reducing odors in privies, increasing pH in stressed digesters, removing phosphorus in advanced wastewater treatment, treating septage, and solids conditioning before and after mechanical dewatering. It is also the principal stabilizing chemical in municipal WWTPs with capacities ranging from 4 to 14 000 L/s (0.1 to 300 mgd) (U.S. EPA, 1979). Larger plants that have used the process include Pittsburgh, Pennsylvania; Memphis, Tennessee; Toledo, Ohio; Blue Plains WWTP in Washington, D.C.; and Clark County Central Plant in Las Vegas, Nevada. According to the U.S. Environmental Protection Agency (U.S. EPA) *Needs Survey of Municipal Wastewater Treatment Facilities* (1989), more than 250 municipal WWTPs use lime stabilization.

Alkaline-stabilized biosolids can be used beneficially in many ways. Each end-use product has particular quality requirements and standards associated with it. Traditional lime stabilization is classified by U.S. EPA as a process to significantly reduce pathogens (PSRP) or Class B requirements stated in the Final Rule *Standards for the Use or Disposal of Sewage Sludge* regulations (U.S. EPA, 1993). Many of the advanced alkaline stabilization technologies meet U.S. EPA's definition of processes to further reduce pathogens (PFRP) and the Class A requirements.

Potential beneficial end-use and disposal options for alkaline-stabilized products include the following:

- Agriculture (organic fertilizer, ag-lime substitute, or soil amendment);
- Nonagricultural land application and reclamation/dedicated land disposal (greater than agronomic application rate);
- Landfills (disposal or beneficial use as daily, intermediate, final, and vegetative cover);
- Manufactured organic topsoil blends;
- Bulk fill applications (slope stabilization, dike construction); and
- Horticulture (nurseries, sod farms).

S TABILIZATION OBJECTIVES

Purposes of alkaline stabilization may include: to substantially reduce the number and prevent the regrowth of pathogenic and odor-producing organisms and, thereby, prevent any health hazard associated with the biosolids; to create a stable product that can be stored; and to reduce the short-term leaching of metals from biosolids not incorporated with natural soil.

The alkaline stabilization process is a simple one. An alkaline chemical is added to raise the pH of the feed, and adequate contact time is provided. At a pH of 12 or higher, with sufficient contact time and homogeneous lime/feed mixing, pathogens and microorganisms are either inactivated or destroyed. Chemical and physical characteristics of the biosolids produced are also altered by the reactions with the alkaline material. The chemistry of the process is not well understood, although it is believed that some complex molecules are split by reactions such as hydrolysis and saponification (Christensen, 1982).

To meet Class B stabilization requirements, the pH of the feed/chemical mixture must be elevated to more than 12.0 for 2 hours and subsequently maintained to more than 11.5 for 22 hours to meet vector attraction reduction by alkaline addition. To meet Class A stabilization, the elevated pH is combined with elevated temperatures (70°C for 30 minutes) or other U.S. EPA-approved time/temperature processes. As long as the pH remains at more than 10 to 10.5, microbial activity and the associated production of odorous gases is greatly reduced or eliminated (U.S. EPA, 1979). However, other odorous gases, including ammonia and trimethylamine, may be produced as a result of the high pH and temperature conditions.

*P*ROCESS DESCRIPTION

There are several different alkaline stabilization technologies available. Each system has advantages and disadvantages; therefore, it is important to evaluate and select a process that will produce the desired product on a site-specific basis. This section describes several alkaline stabilization processes from traditional liquid lime stabilization to newer, emerging dry lime and advanced technologies.

LIQUID LIME (PRELIME) STABILIZATION. Liquid lime or prelime stabilization involves the addition of a lime slurry to liquid sludge to meet Class B stabilization requirements. A typical liquid lime stabilization facility is shown in Figure 6.1. For WWTPs that practice land application of liquid biosolids, such as subsurface injection on agricultural land, lime is added to the thickened biosolids. This practice typically has been limited to smaller WWTPs or those with short haul distances to land application or use.

Solids or septage conditioning with lime before dewatering is a second method of liquid lime stabilization. Lime typically is combined with other conditioners, such as aluminum or iron salts, to achieve enhanced dewatering. This method has been used primarily with vacuum filters and recessed plate

---

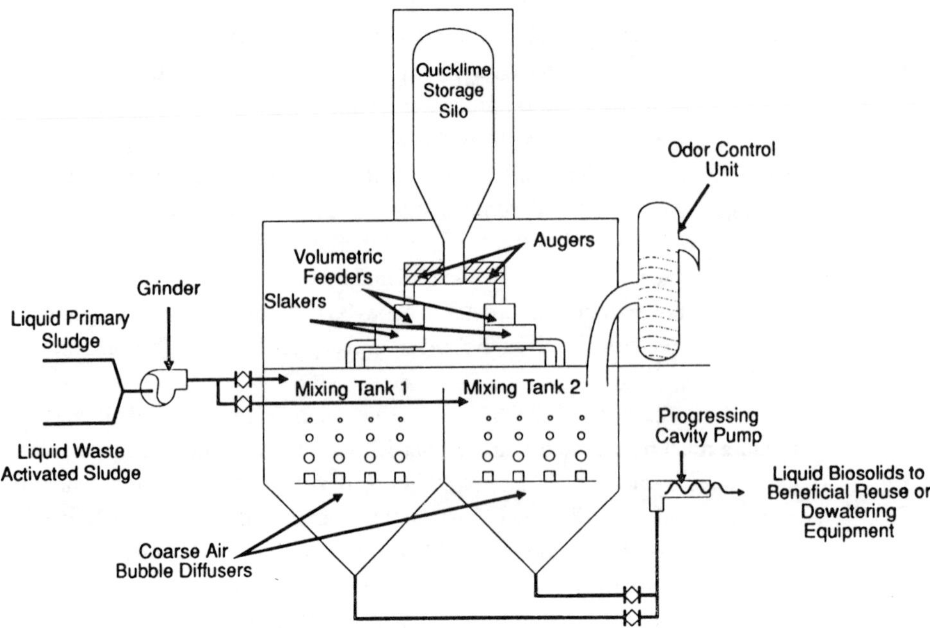

Figure 6.1   Typical liquid lime stabilization facility (U.S. EPA, 1979).

filter presses. Stabilization is complementary in these situations because the lime dose for conditioning typically exceeds the dose required for stabilization.

DRY LIME (POSTLIME) STABILIZATION. Dry lime or postlime stabilization involves the addition of dry quicklime or hydrated lime to dewatered cake. Dry lime stabilization at WWTPs has been practiced since the 1960s (Stone *et al.*, 1992). The lime typically is mixed with the cake using a pug mill, plow blender, paddle mixer, ribbon blender, screw conveyor, or similar device. A process schematic for a typical dry lime stabilization system with pneumatic lime conveyance is shown in Figure 6.2.

Either quicklime, hydrated lime, or other dry alkaline materials can be used for dry lime stabilization, although the use of hydrated lime typically is limited to smaller installations. Quicklime is less expensive and easier to handle than hydrated lime. Additionally, the heat of hydrolysis released in slaking the quicklime when it is added to dewatered cake can enhance pathogen destruction.

The addition of sufficient dry alkaline material can achieve either Class B or Class A requirements as previously defined.

ADVANCED ALKALINE STABILIZATION TECHNOLOGIES. Typical advantages and disadvantages of advanced alkaline stabilization are shown in Table 6.1.

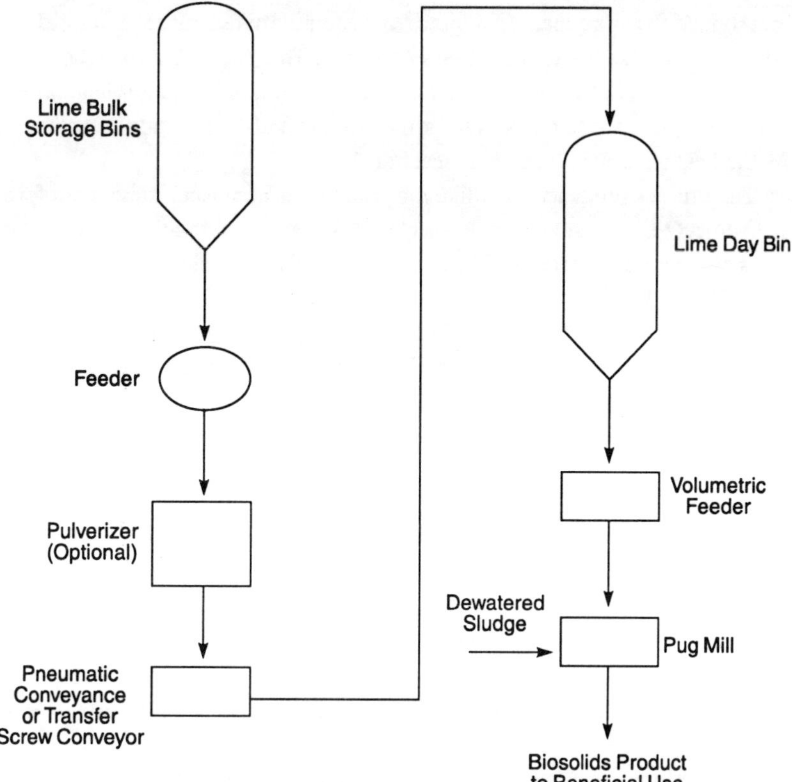

Figure 6.2   Typical dry lime stabilization system process schematic (Oerke and Rogowski, 1990).

New methods of alkaline stabilization using materials other than lime have recently been developed and are being used by municipalities. Most of the technologies that rely on additives such as cement kiln dust (CKD), lime kiln dust (LKD), Portland cement (PC), or fly ash are modifications of traditional dry lime stabilization. The most common modifications include the addition of other chemicals, a higher chemical dose (depending on the chemical), and supplemental drying. These processes alter the characteristics of the feed material and, depending on the process, increase the stability, decrease the odor potential, reduce pathogens, and provide product enhancement.

Many of the processes are proprietary and are available through private firms. The following descriptions are not intended to cover all the available technologies or to rank one process as superior to another. Rather, it is intended to present the scope of processes available to municipalities. A summary of more detailed case study planning, design, and operational considerations on advanced alkaline stabilization processes is available (Engineering–Science, Inc., and Black and Veatch, 1991).

Pasteurization processes use the exothermic reaction of quicklime with water to achieve process temperatures of more than 70°C. This temperature is maintained for more than 30 minutes, as required by federal regulations for add-on pasteurization to meet Class A criteria. This pasteurization reaction must be carried out under carefully controlled and monitored mixing and temperature conditions to ensure uniform treatment and inactivation of pathogens by the heat generated during the reaction.

The process produces a soillike material that is nonviscous and, therefore, not subject to liquefaction under mechanical stress. Varying the process additives and mix ratios produces a range of biosolids-derived materials suitable for use as daily, intermediate, and final landfill cover or land reclamation material (Sloan, 1992). A process schematic for a typical pasteurization process is shown in Figure 6.3. In one variation of this process, pasteurization is carried out in a heated and insulated vessel reactor, where temperatures are maintained at 70°C or above for a minimum of 30 minutes.

A chemical stabilization/fixation technology process typically involves the addition of pozzolanic materials to dewatered cake. The addition of these materials causes cementitious reactions and produces, after drying, a soillike material of approximately 35 to 50% solids content. The only use of this soillike product to date has been as landfill cover material. In many cases, the treated material is further dried at the landfill for 2 to 3 days in small windrows. Class A or PFRP (Oerke and Rogowski, 1990, and Reimers *et al.*, 1981) equivalency has not yet been proven. A typical chemical stabilization process is shown in Figure 6.4.

Another process combines advanced alkaline stabilization with accelerated drying. U.S. EPA has approved two alternative versions of this technology to

Table 6.1 Typical advantages and disadvantages of advanced alkaline stabilization processes.

Advantages	Disadvantages
Meets Class A stabilization requirements.	High annual cost.
Multiple product markets.	High chemical use.
Typically lower capital cost when compared with other Class A stabilization processes.	Extensive odor-control systems required to treat ammonia and other offgases.
	Dewatering facilities required.
Proven with more than 40 installations in the U.S.	Some proprietary processes; annual patent fee could be required
Easy to operate, start up, and shut down	Worker safety concerns with dust from alkaline chemical and ammonia offgas.
Metals concentrations in biosolids are diluted.	Increase on total solids/chemical mass to transport.
Product has value as a liming agent.	Product not appropriate for alkaline soils.
Enclosed facilities for better odor control.	
Properly stabilized product is easy to handle and can be stored in smaller storage facilities.	

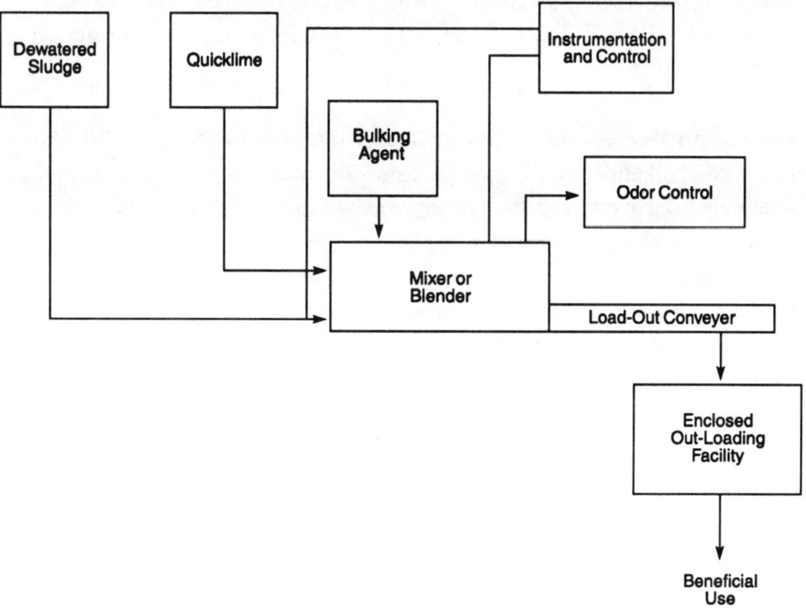

Figure 6.3   Typical pasteurization process schematic.

meet Class A requirements. These alternatives involve the addition of quick-lime, CKD, or other alkaline admixtures and other subsequent processes to create stresses on the pathogens with pH, temperature, ammonia, salts, and dryness (Burnham *et al.*, 1992). One of the alternatives requires an elevated temperature whereby the material is contained for a minimum of 12 hours at 55°C to allow the heat generated by the chemical reaction to further reduce pathogens. The second alternative involves alkaline addition followed by me-chanical drying in windrows to achieve a 50 to 60% solids product. A process

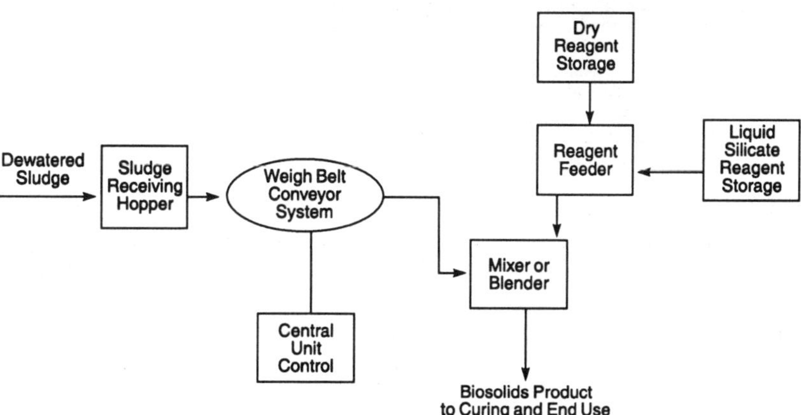

Figure 6.4   Typical chemical stabilization process schematic.

schematic of a typical system is shown in Figure 6.5. The Class A end-use product is used predominantly as an agricultural liming agent, soil conditioner, or as landfill cover.

PROCESS VARIATIONS. Several alternative approaches or modifications to the basic alkaline stabilization process have been developed. Some of these alternatives have evolved from other processes used in a treatment plant; others are associated only with alkaline stabilization.

One example using other processes in the treatment plant to accomplish alkaline stabilization is combining lime-treated primary solids with raw secondary solids for phosphorus removal (Paulsrud and Eikum, 1975). Another example is using existing digesters or other available plant tankage to thicken alkaline-stabilized biosolids before dewatering and disposal (Farrell *et al.*, 1974).

Another alternative associated strictly with the alkaline stabilization process is the use of two mixing vessels: one for initially raising the pH to more than 12, and the other for providing adequate contact time combined with excess lime addition to maintain the pH within the desired range (Counts and Shuckrow, 1975).

Other, nonproprietary processes may also be used to produce Class B alkaline-stabilized biosolids. For instance, Waukegan, Illinois, mixes fly ash and dewatered cake at a 2.0:1 to 2.5:1 ratio. The resultant material is structurally stable and is used to "build" a biosolids-only monofill, as opposed to buying and importing fill material (Byers and Jensen, 1990).

ADVANTAGES AND DISADVANTAGES

Both liquid lime and the dry lime stabilization processes are reliable, low in capital cost, compact, and easier to operate than many other stabilization processes. Many wastewater utilities that use lime stabilization have indicated

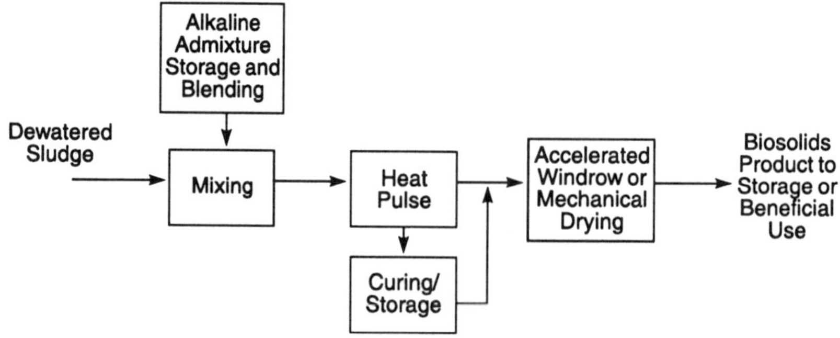

Figure 6.5   Typical alkaline stabilization with subsequent drying process schematic.

that the process greatly reduces odors if homogeneous feed/lime mixing occurs (Kampelmacher and van Noorle Jansen, 1972, and Westphal and Christensen, 1983). Pathogen reduction has been reported to be as effective or better than digestion processes (U.S. EPA, 1979).

Nevertheless, there are disadvantages. The primary disadvantage of alkaline stabilization, as compared with digestion, is that there is no reduction in solids mass. In fact, the mass is increased because of the added lime and chemical formations. Because there is an increase of solids, the amount to be handled is essentially proportional to the dosage rate. The increase in weight typically is greater than the increase in volume; however, there can be a significant loss in total product volume because of temperature associated with lime slaking and the resulting evaporation of water. This increase in mass may increase the costs of transportation for usage or disposal. Any increase in transportation and usage or disposal costs must be weighed against the capital and possible annual savings from using an alkaline stabilization process rather than another process.

The stabilized biosolids can be applied to agricultural land as a source of nitrogen, phosphorus, and beneficial organic matter. The biosolids may also be used to partially or fully replace liming agents for acid soils. The alkaline-stabilized material elevates soil pH and, therefore, restricts the uptake of metals by plants. However, alkaline-stabilized biosolids provide less benefit to the soil and may not be appropriate where the soils are naturally alkaline, as in many parts of the western U.S.

The immobilization of metal ions only occurs to the extent that the pH of the biosolids/chemical mixture is kept high. Also, alkaline-stabilized biosolids typically have lower concentrations of soluble nitrogen and phosphorus on a dry-weight basis than aerobically or anaerobically digested biosolids. Another disadvantage is that homogeneous mixing can be difficult to obtain consistently. In addition, the alkaline stabilization process produces ammonia and possibly other odorous gases that should be treated before exhausted.

*A*PPLICABILITY

Alkaline stabilization has been used in numerous biosolids management programs. Some of the more common situations using alkaline stabilization include the following:

- Short haul distances to end-use or disposal sites—traditional dry lime stabilization is a cost-effective technology for land application or landfill disposal of biosolids. However, because biosolids are not destroyed, it is more cost effective for shorter haul distances.

- Stabilization facilities at small plants—traditional liquid lime stabilization is appropriate at small treatment plants where the small volume of biosolids produced can be readily land-applied. The process is also practical at small plants that store biosolids for later transportation to larger facilities for further treatment or disposal.

- Plants with highly variable solids production—a large portion of the costs for alkaline stabilization are costs for chemicals. Therefore, alkaline stabilization may be cost effective for facilities that only operate seasonally or have several months per year of high production.

- Distribution and marketing programs—advanced alkaline stabilization processes may allow municipalities to operate a distribution and marketing program at a lower capital cost than other technologies such as in-vessel composting or heat drying.

- Backup for existing stabilization facilities—an alkaline stabilization system can be started (or stopped) quickly if the facilities are well maintained and operational. Therefore, it can be used to supplement existing solids treatment capacity or as a substitute for incineration and drying facilities during fuel shortages. Total solids production can be treated with lime when existing facilities are out of service for cleaning or repair.

- Interim solids handling—alkaline stabilization systems have a comparatively low capital cost and, therefore, may be cost effective for plants with short service life.

- Septage disposal—alkaline stabilization is typically used to treat septage and reduce odors before discharge to WWTPs or land application. The U.S. EPA Final Rule *Standards for the Use or Disposal of Sewage Sludge* regulations (U.S. EPA, 1993) require septage to be treated with lime and maintained at a pH of 12 for 30 minutes before land application.

- Supplemental pathogen reduction—alkaline stabilization may be added to processes such as overloaded digesters that have inadequate pathogen reduction; however, strong ammonia odors typically are generated when anaerobically digested biosolids are treated with alkaline materials.

*D*ESIGN CONSIDERATIONS

This section summarizes design considerations for alkaline stabilization processes based on research and on the experiences of pilot- and full-scale facilities using the process.

Because of the interdependency between product quality and process design, the importance of defining process and product goals cannot be overemphasized. A number of design criteria should be evaluated before implementing an alkaline

---

stabilization process. Typical advanced alkaline stabilization design criteria
are shown in Table 6.2. Design criteria will depend on the technology used
and the desired product characteristics, and will vary from site to site. The
typical system design criteria include the following:

- Feed sources and characteristics (quantity, type, quality, percent solids);
- Contact time, pH, and temperature;
- Alkaline chemical types and dosages;
- Solids concentration of the feed/chemical mixture;
- Energy requirements;
- Storage requirements; and
- Pilot-scale testing.

The desired alkaline-stabilized biosolids product is also an important design criterion. A more detailed discussion on product end-use considerations
is included later in this chapter.

FEED CHARACTERISTICS. The amount, sources, and composition of
the feed to be treated determines the overall sizing of the alkaline stabilization facility. Variable performance of thickening or dewatering equipment is
an important consideration when sizing equipment because poor dewatering
performance significantly increases the facility sizing. The solids concentration of the dewatered cake to be stabilized by the alkaline process has an effect on the process because it affects the chemical dosage and facility sizing.
Equipment capacities must be sized to accommodate the volume of feed to be
processed. For example, processing a "wet" cake (10 to 15% solids) will require larger equipment and more alkaline chemical than a drier cake (20 to
25% solids).

The nutrient content of the feed affects the characteristics of the product.
The agronomic benefit of an alkaline-stabilized biosolids product depends on
the amount of plant nutrients contained in the biosolids and the need for a liming agent at the application site. Alkaline stabilization may be advantageous
for untreated solids of marginal agronomic quality because of the dilution (on
a dry-weight basis) of metal pollutants by the alkaline additives.

The type of solids should also be considered when evaluating process design. After being anaerobically digested, biosolids contain 5 to 8 times the
concentration of ammonia–nitrogen, which is converted to a gaseous form
and released at the elevated pH required for the alkaline stabilization process.
Consequently, the potential for odors from ammonia and other nitrogen compounds, such as amines, is increased when anaerobically digested biosolids
are treated by alkaline stabilization. As with all solids processing alternatives,
odor-control facilities typically are required at alkaline stabilization facilities
located near residences or sensitive commercial areas.

Table 6.2   Typical advanced alkaline stabilization design criteria (Fergen, 1991).

Item	Description or equipment	Parameter	Units	Range of value		Selected design value
				Minimum	Maximum	
Materials	Sludge	Solids	percent	20	30	25
		Density	lb/cu ft[a]	45	55	50
	Alkaline bulking chemical	Solids	percent	90	98	95
		Density	lb/cu ft[a]	50	65	65
	Lime	Solids	percent	90	96	95
		Density	lb/cu ft[a]	55	60	60
	Stabilized product	Solids	percent	55	65	60
		Density	lb/cu ft[a]	65	75	75
Curing	Technology	Windrow				
	Detention time	Average	days	3	7	6
		Peak	days	3	7	4
	Temperatures		°C	—	—	52 for 12 hr
	Pile dimensions	Bottom width	ft[b]	6	14	10
		Mix height	ft[b]	2	3	3
		Top width	ft[b]	4	8	6
		Area/unit length	sq ft/ft[c]	10	33	24
		Pile spacing	ft[b]	—	—	5
	Pile turning		lb/d[d]	—	—	1 (typical)
Odor control	Building air	Number of stages	number	1	3	1
		Air changes	number/hr	6	15	12
	Product storage	Number of stages	number	1	3	1
Storage	Sludge	Days of storage	days	0	1	1
	Chemicals	Days of storage	days	5	30	5
	Product	Days of storage	days	80	180	60

[a]   lb/cu ft × 16.02 = kg/m^3.
[b]   ft × 0.304 8 = m.
[c]   sq ft/ft × 0.304 8 = m^2/m.
[d]   lb/d × 0.453 6 = kg/d.

CONTACT TIME, pH, AND TEMPERATURE. Contact time and pH are directly related because the pH must be maintained at the required level for an adequate time to destroy pathogens. The chemical added must provide enough residual alkalinity to maintain a high pH until used or discarded. The high pH will prevent growth or reactivation of odor-producing and pathogenic organisms.

The drop in pH (referred to as pH decay) occurs in the following sequence. Atmospheric carbon dioxide or acid rain (which forms a weak acid when dissolved in water) is absorbed, then gradually consumes the mixture's residual alkalinity, and the pH gradually decreases. Eventually, the pH drops to less than 11.0 and renewed bacterial action will resume and the renewed production of organic acids will cause the pH decay to continue (similar to the reactions within the anaerobic digestion process).

The pH typically drops during the stabilization process and, therefore, should be raised to more than 12 and maintained at that value. Biosolids do not have to be contained within a contact vessel as long as the pH can be monitored to ensure that it remains at the desired value for the desired period.

ALKALINE CHEMICAL TYPES AND DOSAGES. The types and dosages of the alkaline chemicals to be added are important design criteria. The quality of the chemicals (lime, CKD, PC, and LKD) should be consistent. It has been found that different types or sources of additives result in different product texture and granularity. Lime is available from numerous sources, ranging from a high calcium lime obtained from oyster or clam shells to a relatively low calcium dolomitic lime. Major considerations in selecting the type of chemical to be used include economics, availability, desired mixing, and desired end-product characteristics. Table 6.3 lists typical alkaline chemical characteristics.

Some alkaline reagents are considered byproducts from other industries (such as CKD or LKD), and it is important to ensure that this material does not introduce contaminants or additional pollutants, jeopardizing the quality of the product. Cement kiln dust from hazardous waste kilns should be avoided. A quality assurance/quality control program with frequent sampling and analysis should be developed to ensure consistent quality of the biosolids product.

Because the quality of the alkaline additives may have a direct effect on the quality of the product, adequate monitoring and proper management are important to ensure consistent quality. Because the characteristics of a byproduct such as LKD, CKD, or fly ash can vary from one location to the next, consistent vendor quality control procedures are essential. The material from a single kiln or furnace will remain fairly consistent, provided that operating conditions do not change drastically. More importantly, pilot- or bench-scale testing should be performed to determine how variations in the alkaline

---

Table 6.3   Characteristics of common alkaline materials (Lewis and Gutschick, 1980).

Common name formula	Available forms	Appearance and properties	Bulk density	Commercial strength	Referenced standards
Quicklime/CaO	Pebble; crushed; lump; ground; pulverized (at least 80% passes 100 sieve)	White (light gray, tan) lumps to powder; unstable, caustic irritant; slakes to hydroxide slurry evolving heat 1.1 MJ/kg (490 Btu/lb); reacts with CO_2 in air to form $CaCO_3$; saturation solubility approximately pH 12.5	880 to 1 200 kg/m^3 (55 to 75 lb/cu ft); specific gravity 3.2 to 3.4	70 to 96% CaO (below 88% can be poor quality)	ASTM C911, FHWA/RD-82/167
Hydrated lime/Ca(OH)$_2$	Powder (at least 75% passes 200 sieve)	White, 200 to 400 mesh; powder free of lumps; caustic, dusty irritant; absorbs H_2O and CO_2 from air to form $Ca(HCO_3)_2$; saturation solubility approximately pH 12.4	400 to 640 kg/m^3 (25 to 40 lb/cu ft); specific gravity 2.3 to 2.4	$Ca(OH)_2$—82 to 98%; CaO—62 to 74% (standard 70%)	ASTM C911, FHWA/RD-82/167
Cement kiln dust	Powder (at least 75% passes 100 sieve; at least 50% passes 200 sieve)	Light gray to brown; properties variable, depending on kiln and operating conditions	50 to 75 lb/cu ft[a]	Variable	ASTM C150, ASTM C911, FHWA/RD-82/167
Lime kiln dust	Powder (at least 75% passes 100 sieve; at least 50% passes 200 sieve)	Light gray to tan; properties variable, depending on kiln and operating conditions	50 to 75 lb/cu ft	Variable	ASTM C150, ASTM C911, FHWA/RD-82/167

Table 6.3   Characteristics of common alkaline materials (Lewis and Gutschick, 1980) (continued).

Common name formula	Available forms	Appearance and properties	Bulk density	Commercial strength	Referenced standards
Portland cement	Powder; Type I cement is typically used	Light gray	N/A[b]	N/A	ASTM C150
Fly ash	Powder (at least 70% passes 200 sieve)	Variable	Variable	Variable	ASTM C593

[a] lb/cu ft $\times$ 16.02 = kg/m^3.
[b] N/A = not applicable.

additives will affect product quality and how the process and chemical dosages should be adjusted to compensate for these variations.

There are two predominant types of lime available: quicklime (calcium oxide) and calcium hydroxide. Slaked lime in a liquid slurry (carbide lime) is also available. Carbide lime is a byproduct of the manufacture of welding-grade acetylene from calcium carbide. The application principles of this material are the same as those for either calcium hydroxide or quicklime in slurry form; therefore, carbide lime is not specifically discussed. The decision of which type of lime to use should be based on economics and materials handling characteristics (such as alkaline material particle size). The production and transportation costs of calcium hydroxide are approximately 30% more than the costs of quicklime. However, for liquid operations, it has the advantage of requiring less equipment at the facility because it has already been hydrated (slaked). Typically, calcium hydroxide is economical for use at small facilities; however, when use exceeds 2.5 to 3.5 Mg/d (3 to 4 ton/d), quicklime should be considered.

Aside from the cost, the primary advantage of calcium hydroxide over quicklime is the elimination of slaking equipment at the facility. If quicklime is added directly to dewatered cake (dry lime stabilization), slaking is not necessary; however, the additional handling precautions that must be addressed because of the exothermic reaction of quicklime and water should be evaluated. Another advantage of dry lime stabilization is the elimination of lime sidestreams and the related abrasion and scaling of piping and mechanical equipment.

The required dosages of the specific chemicals will depend on the raw feed type (primary, waste activated, trickling filter, or septage) and quality, chemical composition (including organic content) of the feed and liquid, the

Table 6.4   Lime dosage required for liquid lime stabilization at Lebanon, Ohio[a] (U.S. EPA, 1979).

Type of sludge	Average solids concentration, %	Average lime dosage, lb calcium hydroxide/lb dry solids	Average pH	
			Initial	Final
Primary sludge[b]	4.3	0.12	6.7	12.7
Waste-activated sludge	1.3	0.30	7.1	12.6
Anaerobically digested combined	5.5	0.19	7.2	12.4

[a] Dose required to maintain pH 12 for 30 minutes.

[b] Includes waste-activated sludge.

feed solids concentration, the desired product, and the type and quality of the alkaline material.

Table 6.4 shows the range of liquid lime stabilization dosages required to maintain a pH of 12 for 30 minutes (U.S. EPA, 1979). Numerous researchers have confirmed these dosages (Ramirez and Malina, 1980).

Chemical composition of the feed also determines the required chemical dose. This composition depends on the type of solids and the process used (for example, chemical coagulation) for treatment. Another factor that affects chemical dose is solids concentration. Figure 6.6 displays the typical relationship found at municipal plants between liquid lime dosage and pH for a typical feed at several solids concentrations (U.S. EPA, 1975). Table 6.5 shows a wide range of lime doses from 10 to 60% on a dry-weight basis. As the percent of solids concentration increases, the required dose typically increases. The required dose per unit mass of solids tends to be somewhat higher for dilute feeds (less than 2.0% solids), because more lime is required to raise the pH of water. However, liquid lime requirements are more closely related to

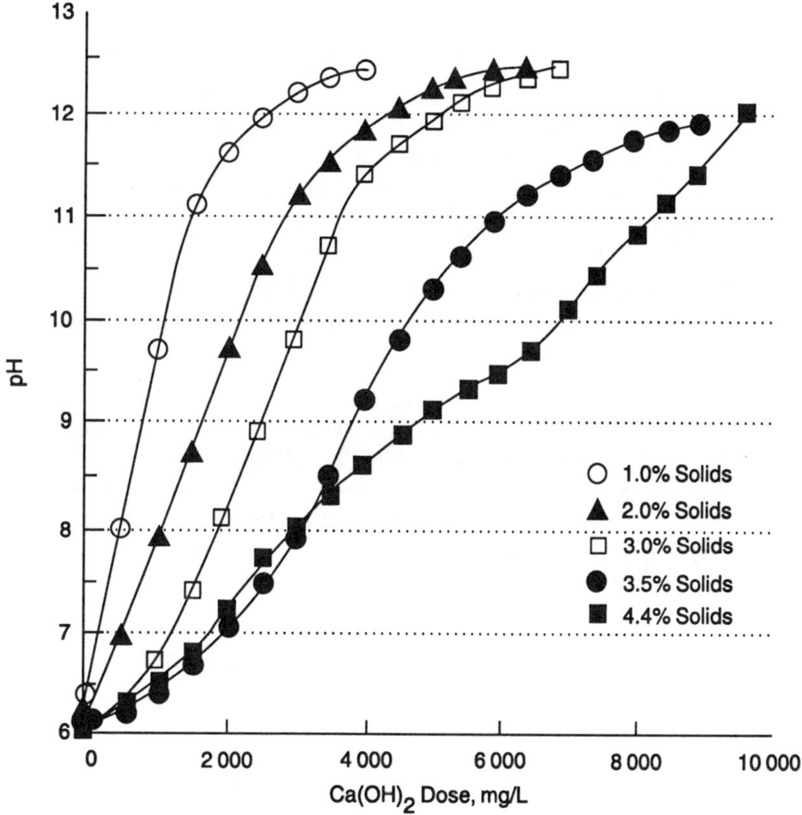

Figure 6.6 **Liquid lime stabilization doses required to raise pH in a mixture of primary solids and trickling filter humus at different solids concentrations (U.S. EPA, 1975).**

the total mass of solids than to volume over a range of solids concentrations typical in wastewater treatment (0.5 to 4.5%) (U.S. EPA, 1979). Volume reduction by thickening may have little or no effect on the amount of lime required because the solids mass is not significantly changed.

Minimum typical lime doses of 25 to 40% on a dry-weight basis as calcium hydroxide are required for liquid lime Class B stabilization before vacuum filtration (Figure 6.7). The curves on Figures 6.7, 6.8, and 6.9 show the characteristic pH drop that occurs when inadequate liquid lime is added. When the dose is too low, the pH of the feed-lime mixture may initially reach the target value of 12, but a rapid pH decay may follow.

Typically, lower minimum doses of 15 to 30% on a dry-weight basis as calcium hydroxide are required for effective dry lime versus liquid lime stabilization. Figure 6.10 shows a wide range of adequate dry lime Class B stabilization doses of 13 to 40% as calcium hydroxide on a dry-weight basis.

Figure 6.11 shows the theoretical dry lime stabilization dose for both Class B and Class A stabilization. The lower line shows maximum pH requirements, and the upper line shows Class A temperature requirements. Figure 6.11 is based on a quicklime dose requirement of 25% on a dry-weight basis. Note that while the quicklime requirement for Class B stabilization theoretically increases with increased solids, the quicklime requirement for Class A stabilization decreases with increased solids. This is because of the predominance of the heating requirements of the water for Class A versus the pH effects of Class B (*Municipal Sewage Sludge*, 1992).

The following assumptions were used for the Class A temperature requirements in Figure 6.11:

- That feed input temperature equals 20°C (68°F).
- That 100% of the quicklime reacts with water in the feed to produce heat (1140 kJ/kg, or 490 Btu/lb of quicklime).

Table 6.5   Liquid lime stabilization doses required to keep pH above 11.0 for at least 14 days (Farrell *et al.*, 1974).

Type of raw sludge	Lime dose, lb calcium hydroxide/lb suspended solids[a]
Primary sludge	0.10 – 0.15
Activated sludge	0.30 – 0.50
Septage	0.10 – 0.30
Alum sludge[b]	0.40 – 0.60
Alum sludge[b] plus primary sludge[c]	0.25 – 0.40
Iron sludge[b]	0.35 – 0.60

[a] lb/lb × 1 000 = g/kg.

[b] Precipitation of primary treated effluent.

[c] Dry-weight basis.

- That quicklime is 100% calcium oxide; typically, this value is 90%.
- That the specific heat of the feed is 0.25.
- That no heat loss from the feed to the air or the equipment occurs.

In practice, these conditions rarely exist; therefore, the amount of quicklime used to achieve Class A can be up to 50% greater than the Class A temperature line (top) in Figure 6.11. A drier, more easily crumbled product can be achieved by increasing the quicklime addition up to as much as twice the theoretical value shown.

Chemical dosages for advanced alkaline stabilization technologies vary by process, types of chemical, and product requirements. A typical material balance for an advanced alkaline stabilization facility assuming a 65% chemical dose on a wet-weight basis is shown in Table 6.6. Note that a lime dose expressed on a wet-weight basis is four times greater than a dose expressed on a dry-weight basis for a dewatered cake material with a 25% solids concentration. For example, the chemical dose in Table 6.6 of 65% on a wet-weight

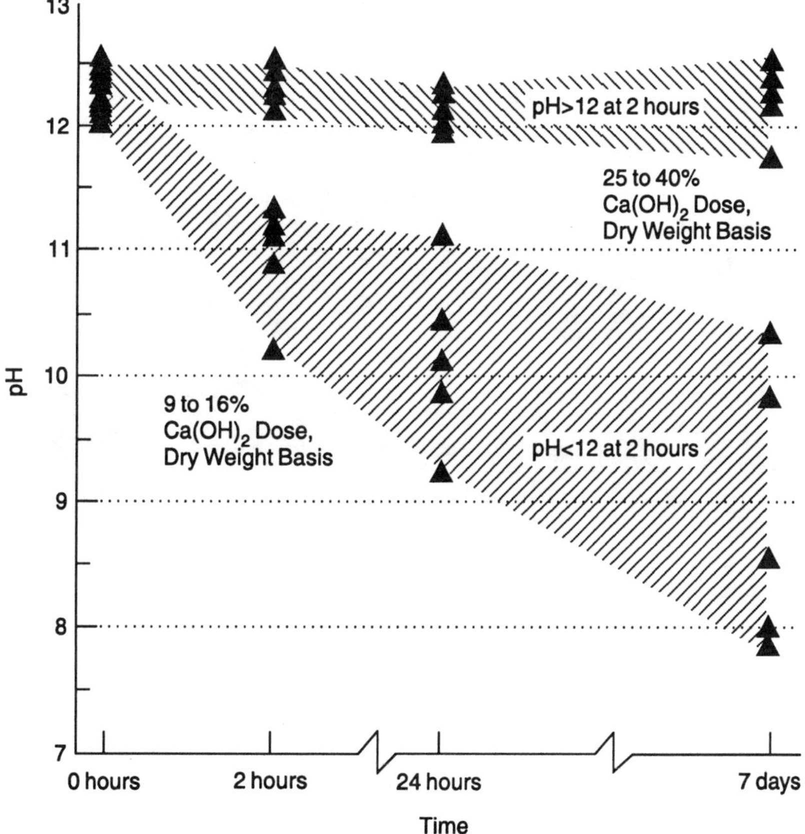

Figure 6.7   **Example of pH decay following liquid lime stabilization before vacuum filtration (Westphal and Christensen, 1983).**

basis is equal to approximately 245% on a dry-weight basis. Material balances should be used to size alkaline stabilization facilities and determine initial and product characteristics.

The data presented in Tables 6.2, 6.4, 6.5, and 6.6 can be used for preliminary design of liquid and dry lime stabilization facilities; however, it is recommended that the required dosage be determined on a case-by-case basis because of the many factors that affect this dosage (Farrell *et al.*, 1974). To prevent pH decay and the associated regrowth of organisms, the lime dose may have to be increased beyond that necessary for stabilization (Ramirez and Malina, 1980). The exact dosage for any particular feed can be estimated through laboratory testing.

SOLIDS CONCENTRATION OF FEED/CHEMICAL MIXTURE. The solids concentration of the feed/chemical mixture is an important design consideration for materials handling purposes. Minimum solids concentrations may be required to meet regulatory requirements such as for landfilling or ex-

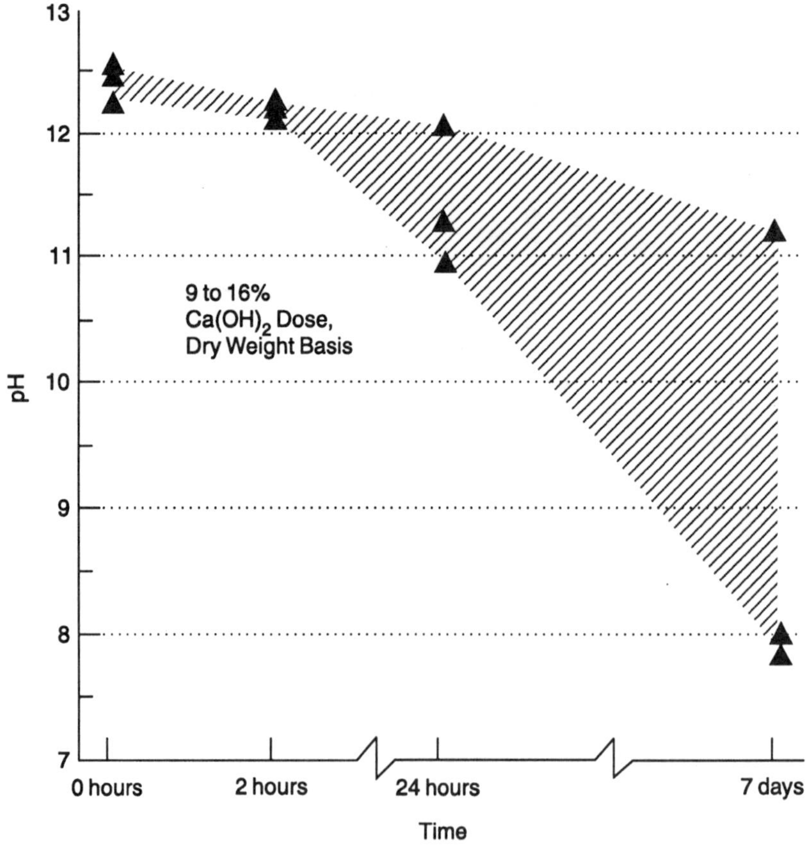

Figure 6.8 Example of pH decay following liquid lime stabilization (Westphal and Christensen, 1983).

tended storage. Solids concentration (dryness) and granularity will affect the materials-handling characteristics for trucking and application/disposal equipment.

The solids concentration of the initial feed/chemical mixture also affects any supplemental drying step in advanced alkaline stabilization processes. Through the addition of the alkaline additive, chemical reactions occur that increase the apparent solids content of the mix. This increase in solids is because of the cumulative consequence of the addition of solids, the chemical binding, and evaporation of the water in the feed. The alkaline material, particularly quicklime, produces a fast reaction resulting in a temperature increase within a matter of minutes. Thorough mixing of the feed and alkaline material is important to achieve the target solids content and pathogen destruction and to reduce residual odors (such as ammonia) in the product.

By using a high chemical dose, it is possible to achieve the desired solids concentration, thereby reducing or eliminating the need for supplemental drying. However, excessive chemical costs may make this practice prohibitively expensive. It is possible to add other bulking materials to increase the product dryness and improve handling characteristics but not increase the chemical dosage. Potential bulking materials include fly ash, wood ash, sawdust, sand,

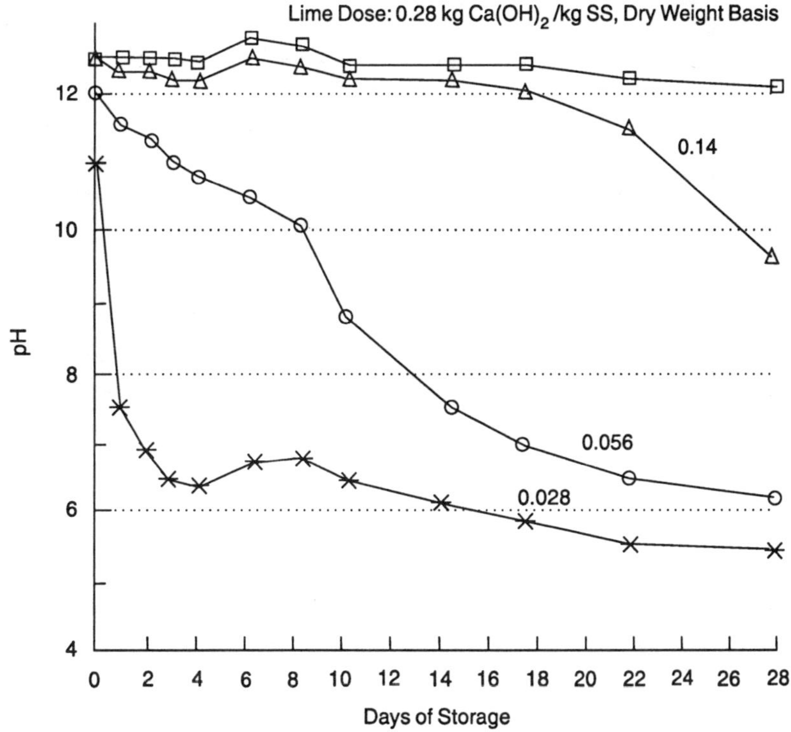

Figure 6.9　Change in pH during storage of raw primary solids using different liquid lime stabilization dosages (Farrell *et al.*, 1974).

and soil. In addition, the quality of the product may be affected. Mechanical mixing in a windrow operation enhances drying, blends the material, and releases trapped ammonia and other volatile gases created during dewatering operations, resulting in a more homogeneous product. The final design should reflect the best balance between the chemical dose and the amount of subsequent drying required.

ENERGY REQUIREMENTS. During the liquid lime stabilization process, the principal energy use is for mixing the liquid sludge and lime slurry. For dry alkaline stabilization, mixing energy requirements are minimal and vary with throughput, chemical dose, and mixer type.

Transport vehicles (feed, chemicals, and product end-use) and air ventilation and scrubbing equipment for ammonia and odor-control can also be substantial energy users.

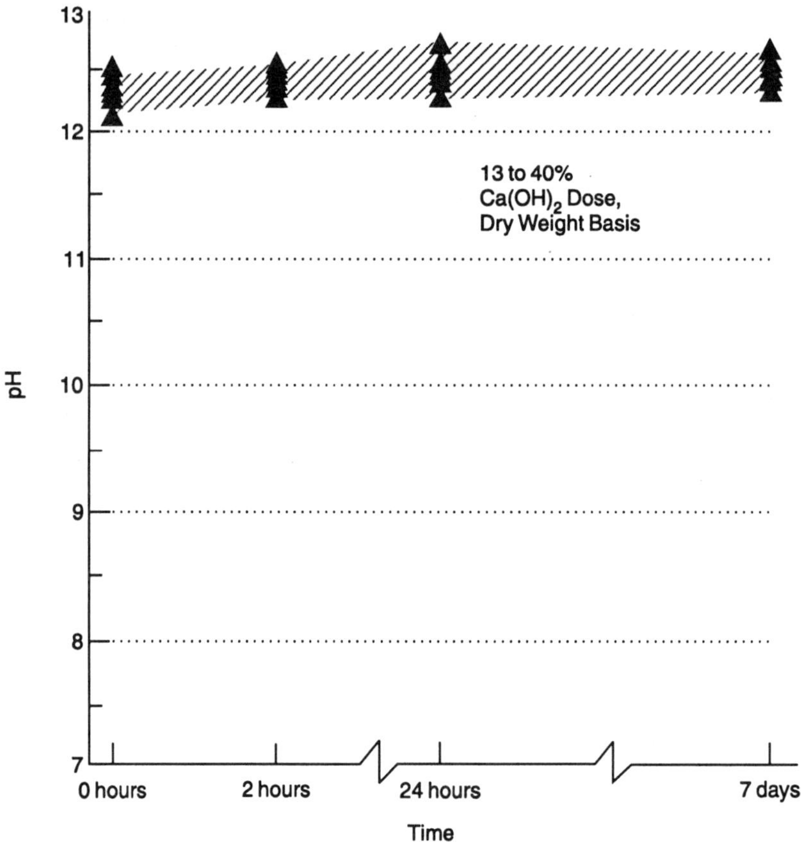

Figure 6.10   Example of pH decay following dry lime stabilization of raw primary/waste activated dewatered cake (Westphal and Christensen, 1983).

STORAGE REQUIREMENTS. The size of storage facilities should be tailored to the actual needs of each facility and should include facilities for intermediate storage and product storage.

Intermediate Storage. Some advanced alkaline stabilization processes require additional facilities for the heating step to achieve Class A stabilization requirements. The objective of using this step is to contain the heat produced from the exothermic reaction, reducing the chemical dosage by reducing heat loss. Intermediate storage facilities can include insulated steel, live bottom hoppers, concrete bunkers, or an uninsulated stockpile in an open concrete pad.

Product Storage. Final storage of the alkaline-stabilized biosolids product is another important design consideration. Adequate storage capacity is required where markets are seasonal or have not been established. The amount of storage required depends on the type of end-use product or the distribution and

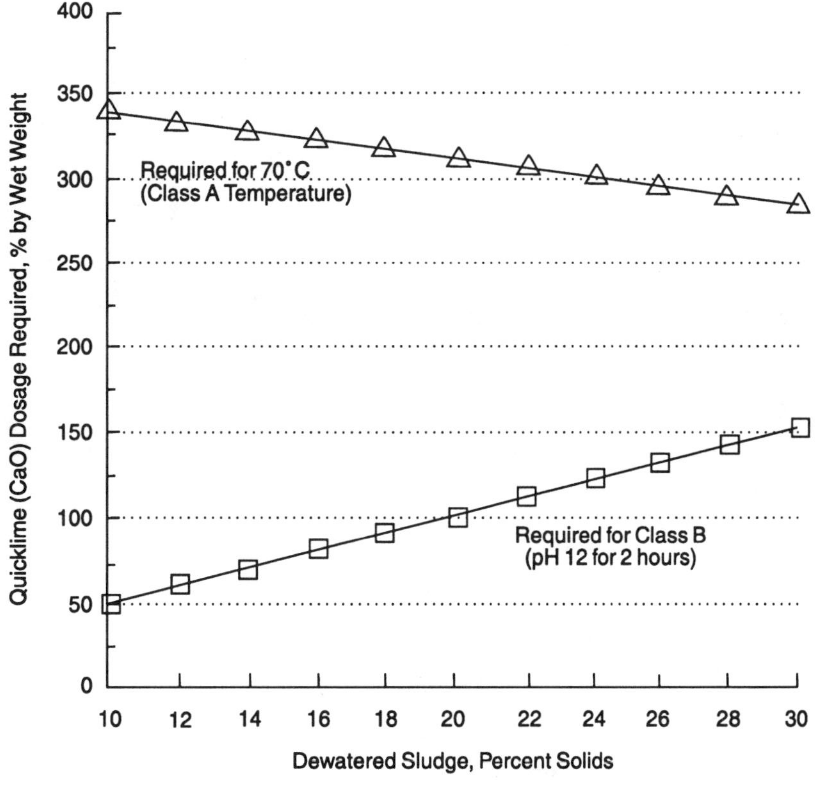

Figure 6.11 **Theoretical quicklime requirements for Class B and Class A dry lime stabilization processes at varying solids concentrations (*Municipal Sewage Sludge*, 1992).**

Table 6.6   Typical materials balance for advanced alkaline stabilization facilities (Fergen, 1991).[a]

Process	Item	Solids content, %	Volume, cu ft[b]	Total weight, ton	Dry weight, ton	Bulk density, lb/cu ft[c]
			Solids balance			
Mixing	Sludge cake	25.0	6 400	160.0	40.0	50.0
	Chemicals	95.0	3 186	103.5	98.4	65.0
	Initial mix	52.5	9 586	263.5	138.4	55.0
Windrow	Initial mix	52.5	9 586	263.5	138.4	55.0
	Evaporation loss	—	3 437	32.9	—	—
	Product	60.0	6 149	230.6	138.4	75.0

Chemical dose

Wet-weight basis 65%
Dry-weight basis 245%

[a] For peak conditions multiply all the values by peaking factor (except density and percent solids).
[b] cu ft × 28.32 = L.
[c] lb/cu ft × 16.02 = kg/m^3.

marketing methods. Sufficient storage (typically a minimum of 30 to
90 days) should be provided if product curing is required and to accommodate road and weather conditions and fluctuations in the product marketing
and distribution schedule. If the material is to be used as landfill cover, storage requirements will most likely be minimal. Weekend storage is needed if
the landfill operates only 5 days per week. Several alkaline stabilization facilities are currently pursuing markets for their stabilized products: an agricultural fertilizer/liming agent and an agricultural soil amendment. Until the
markets have been established, the products must be stockpiled or discarded.
Because the demand for these agricultural products is seasonal, the facility
must have provisions to stockpile during periods of low demand.

Product solids concentration and long-term product stability are important
considerations when designing product storage facilities. The product storage
facilities should be sized to meet the actual needs at each facility. A typical
product storage period is 30 to 90 days. Provisions must also be included for
storage during scheduled and unscheduled equipment maintenance if maintenance includes downtime. The storage facility should be designed to prevent
deterioration of product quality during inclement weather. In many climates,
it may be desirable to provide covered storage. Where uncovered storage is
used, provisions should be included for leachate and runoff collection to
avoid ponding and, in some instances, treatment. Runoff from stockpiles of alkaline-stabilized biosolids can result in stagnation and septic odors (Engineering–
Science, Inc., and Black and Veatch, 1991).

PILOT-SCALE TESTING. Because quality and consistency of feed vary
from site to site, qualitative and quantitative analyses must be performed to
determine the specific chemical dosages and process design parameters.
Pilot-scale testing should be used to determine the optimum chemical dosages and mixer performance. Pilot-scale testing also allows the municipality
to evaluate different operating procedures, as well as end-use products.

Bench- and pilot-scale testing should be conducted before implementation
of an alkaline stabilization process. Four primary areas to be evaluated include

- Process requirements, including alkaline material types and dosages;
- Equipment, including energy requirements;
- End-use product quality, including desired product solids concentra-
  tion and granularity; and
- Generation and control of odors.

Process concerns include the chemical types and dosages, feed and product solids concentration, and other process steps such as supplemental heating
and drying, as required. The chemical dose must be determined to meet the
pH, solids content, heat rise, and end-use product requirements. The chemical
dosages can be estimated in bench-scale tests using carefully measured

volumes of feed mixed with varying dosages of chemicals. All pilot-scale testing should include generation of mass balance calculations to ensure consistency between chemical dosages and product solids concentration. Full-scale pilot equipment should be used, if possible, to assess actual chemical dosage rates and mixing performance. It is extremely important that the testing conditions be controlled to simulate field conditions to the greatest extent possible. For example, during the winter, a different chemical dose or a modification of the chemical formula may be required to achieve the desired end-use product. In large-scale windrow drying operations, carbon dioxide mixed into the product can lower the pH. Therefore, the dosage of alkaline material may have to be increased to maintain the proper pH. Samples should be cured in closed or open containers of the type that will be used in the full-scale system.

The initial solids concentration of the feed/chemical mixture should be tested for compatibility with the proposed drying technique. It may also be useful to investigate various chemical dosages in different drying/curing configurations. The chemical dose can have a significant effect on the rate of drying and the corresponding drying area requirements.

Pilot-scale testing of equipment may be considered to evaluate equipment requirements. The goal of the testing is to determine the equipment, energy, and chemical requirements for producing the desired product compatible with the end-use or next processing step. It was found that use of an inappropriate paddle configuration or operating speed on a pug mill feed/chemical mixer at a dry lime stabilization facility resulted in an undesirable material. Proper mixing is necessary not only to achieve the desired product characteristics but also to ensure thorough blending of the alkaline additive.

Other process parameters that should be considered during pilot- or bench-scale testing include emission of odors from the process and product; levels of plant nutrients, metals, and organic chemicals; and compatibility of the alkaline material dose with the dewatering polymer. Some polymers may deteriorate in high alkaline conditions and exhibit strong trimethylamine "dead fish" odors (Jacobs and Silver, 1990). Various dosages of alkaline material should be tested with different polymers to determine the effects on odor release and physical characteristics of the products (such as compaction and granularity).

The final item evaluated by pilot-scale testing is the product. Pilot- and bench-scale testing provide excellent opportunities to investigate product quality and marketability before beginning full-scale production. It is helpful to invite prospective users to observe pilot-scale tests or small-scale demonstration programs to encourage interest in the product. Physical characteristics, such as solids content, pH decay, leachability, permeability, or unconfined compressive strength, should be evaluated if the product will be used for landfill or slope stabilization. The quality of the product should also be tested to provide the data and documentation required for regulatory approval.

*D*ESCRIPTION OF PHYSICAL FACILITIES

SOLIDS HANDLING AND FEED EQUIPMENT. Dewatered cake handling equipment consists chiefly of belt and screw conveyors and pumps. Belt conveyors typically are used on horizontal or gently sloping planes. Common belt conveyor problems, such as minor spillage, slippage, and frequent bearing maintenance, have occurred at many facilities. Screw conveyors also are used to transfer dewatered cake to the alkaline stabilization process mixer or storage hopper. Screw conveyors and high-pressure dewatered cake pumps can physically "condition" the cake and make it difficult and sometimes impossible to mix it homogeneously with a dry alkaline chemical. Some screw conveyors have a tendency to roll the cake/chemical mixture into "balls." Pumps can compact the dewatered cake into a long tube that requires breaking at the mixing step. The rolled "balls" and compacted cake can be especially critical when dealing with the product, being either desirable or undesirable depending on the final objective.

Although the alkaline stabilization processes are relatively simple, a regular inspection and maintenance program is essential. The conveying system and other moving parts must be closely monitored for wear. If only one conveyor is used to feed the alkaline stabilization process, it is important that the equipment be routinely inspected, maintained, and calibrated. Conveying system downtime could delay or halt the stabilization process. If multiple process trains are used, bypasses and crossovers should be provided to avoid excessive downtime.

Redundant storage and processing equipment trains should be seriously considered to allow routine maintenance and calibration and operational flexibility without downtime. While perhaps not as desirable as dual-operating systems, temporary portable units can be placed into operation in a matter of hours or days, if necessary. Storage hoppers/bunkers between the dewatering system and the alkaline stabilization system may be used to dampen variations in dewatering system output and to allow each process to operate independently.

ALKALINE MATERIAL STORAGE AND FEEDING. The alkaline stabilization process requires special chemical storage and feeding equipment. Traditionally, alkaline chemical storage should include provisions to meet a minimum of 7 days' demand, with a 2- to 3-week supply preferred. Calcium hydroxide can be stored for up to 1 year. Quicklime deteriorates more rapidly and should not be stored longer than 3 to 6 months. The absolute minimum storage capacity recommended is 200% of the bulk chemical shipment volume, depending on the distance between the chemical supplier and the user.

Because of the high chemical use associated with some advanced alkaline stabilization processes, use of traditional design criteria can result in excessively large storage capacity; however, smaller capacity (2 to 3 days of chemical use) can be considered to reduce capital costs. Consequently, reliable chemical delivery arrangements are imperative. If a chemical delivery is delayed or missed, the facility may not be able to operate. The costs associated with daily chemical delivery should be compared with the costs of extra storage capacity. If the source of chemical supply is located near the facility, extra storage capacity may not be needed. However, if a distant chemical supplier is selected, extra storage capacity may be advisable to ensure continuous operation.

Quicklime can be stored in lump or pebble form and ground on site to reduce the potential reaction with moisture during storage, especially if the alkaline material is to be stored for up to 6 months.

Alkaline material is stored in steel silos with hoppers that have a side slope of at least 60 deg. Bulk storage silos and day chemical bins, if used, should be equipped with dust collectors and live bottom bins, hopper agitation, or air pads to facilitate unloading and reduce clogging or bridging.

With any type of chemical, there are potential problems. During storage, lime can react with carbon dioxide in the air to form a calcium carbonate coating on the surface of the lime particles, making the lime less reactive. Quicklime and other alkaline materials react readily with moisture from the air, which causes caking that may interfere with feeding and slaking. Therefore, lime should be stored in dry facilities and protected against moisture to prevent accidental slaking. Because the slaking process generates heat, quicklime should not be stored near combustible materials.

Dry alkaline materials can be conveyed mechanically with a screw conveyor if the distance from the bulk storage silo to the chemical addition point is short. Dry alkaline materials can also be conveyed pneumatically, either under pressure or under a vacuum. Each type of pneumatic conveyance has its benefits. Dust problems are reduced under a vacuum because any leaks are into the system. The pressure systems can move larger volumes of material. Pneumatic conveying air should be predried to reduce hydration and other moisture-related problems; however, pneumatic transfer systems may have problems maintaining homogeneous chemical bulk densities if a variety of alkaline materials are used (Rubin, 1991).

A wide variety of chemical feed equipment is available. Chemical metering equipment includes volumetric screw feeders, rotary airlock feeders, and gravimetric feeders. The volumetric feeder delivers a constant volume of alkaline material regardless of its density. The gravimetric feeder delivers a constant mass of alkaline material and provides more accurate control. However, the gravimetric feeder costs approximately twice as much as the volumetric type. An evaluation should be performed to determine the appropriate feeder for each particular application (Rubin, 1991). Other reported chemical feed problems include jamming by foreign objects, such as wire or rocks. The feed equipment should be isolated from the storage silo with a slide gate or similar device. This allows easy removal of the metering equipment if it becomes jammed.

Dust problems have also been experienced with most of the chemical feed systems. Poorly fitting slide gates and leaking feeders are obvious sources of dust. The vertical drop between the feeding equipment and the process mixers should be reduced or enclosed to reduce dust problems.

Because the powdery PC is hygroscopic, it tends to become packed in the corners of the storage hopper. The backup of PC in pipes is caused primarily by the moisture generated during the mixing process in the pug mill.

Moisture can be generated in the mixing step with feed and PC or any alkaline chemical and may also rise into the chemical feed and storage equipment. Venting the mixing device away from chemical feed and storage equipment can reduce this operational problem.

LIQUID LIME CHEMICAL HANDLING AND MIXING REQUIREMENTS. Lime typically is fed to liquid sludge in the slurry ("milk of lime") form. Dry lime cannot be effectively added to liquid sludge because caking will occur.

After being mixed into a slurry, both calcium hydroxide and slaked quicklime are chemically the same, and the same feeding processes can be used for both. Lime slurry can be prepared by either the batch or the continuous method. Batch slurry consists of dumping the bagged lime into a mixing tank. The minimum power requirement for mechanical mixing of calcium hydroxide slurry at a concentration of 120 kg/m^3 is approximately 200 kW/m^3 to ensure initial wetting and dispersion (Beals, 1976). The slurry is then metered into the mixing tank for mixing and contact.

The contents of slurry tanks are agitated by compressed air, water jets, or mechanical mixers. The slurry is then transferred to the mixing tank as needed. This transferring process may be the most troublesome operation in the facility. The slurry reacts with bicarbonate alkalinity in the makeup water and with atmospheric carbon dioxide to form calcium carbonate scale that can plug lines. The magnitude of this problem increases with increasing transfer distances and with the amount of bicarbonate or carbon dioxide in contact with the slurry. Therefore, the slurry tanks should be located as close to the mixing basin as possible. Cascading weirs and other facilities that cause turbulence should be avoided.

The basic difference between using quicklime and calcium hydroxide to stabilize liquid sludge is that slaking equipment must be used with quicklime. Slaking can be done either on a batch or continuous basis. The batch method is more appropriate for small-scale facilities; however, the use of quicklime is typically less advantageous for small facilities. The slaking process consists of mixing one part of quicklime with approximately two to three parts water to create a lime paste. This paste should be held for approximately 30 minutes to allow complete hydration in the slaking chamber. The hydration reaction is exothermic (heat is released). Elevated temperature is necessary for proper slaking. However, localized boiling and spattering may occur, creating a potentially hazardous condition. After slaking, the paste enters a dilution chamber where it is diluted to the desired concentration and grit is removed.

The appropriate type of automatic equipment for continuous slaking depends largely on the proportion of lime to water. This proportion depends on both the type of lime and the equipment used. Therefore, both the type of lime and the equipment should be considered when designing a slaking system.

A stabilization tank is recommended downstream of the slaker to ensure that all chemical reactions between the calcium hydroxide and the dissolved solids in the water have been completed. This reduces scaling in downstream portions of the system. Lime slurry from the slakers should discharge directly to the stabilization tank, if possible. The detention time in the stabilization tank should be at least 15 minutes. Adequate mixing is required to maintain particles in suspension and to prevent short circuiting. For cylindrical tanks with a 1:1 height-to-diameter ratio, minimum horsepower requirements are approximately as shown in Table 6.7.

Consultation with agitation equipment suppliers is recommended. Baffles may be required to prevent vortex formation and should be designed to prevent solids buildup in the corners depending on tank geometry.

A cleaning system using dilute hydrochloric acid should be provided to remove calcium carbonate scale from pumps and piping. Consequently, the materials used in the pumps and piping materials must be compatible with both an acid and a caustic environment.

To facilitate scale removal, flexible piping or open troughs should be used to convey lime slurry whenever possible. Lime slurry may be abrasive, particularly if low-grade pebble lime is used, and materials and equipment should be selected accordingly.

The primary purpose of the mixing tank is to provide adequate mixing and contact time of the raw feed with the lime slurry. The recommended contact time is approximately 30 minutes after the pH reaches 12.5. The mixing time

Table 6.7   Minimum mixing horsepower requirements for liquid sludge and lime slurries (WPCF, 1988).

Slurry concentration, lb/gal[a]	hp/1 000 gal[b]
1	0.25
2	0.50
3	1.00

[a] lb/gal × 0.119 8 = kg/L.

[b] hp/1 000 gal × 197.3 = kW/m^3.

criterion is site-specific, and bench- or pilot-scale testing should be done whenever possible.

The tank can be constructed of mild steel. The size of the tank will depend on whether mixing will be done on a batch or a continuous basis.

Batch mixing tanks typically are used for smaller facilities. The tanks should be sized to treat the daily solids production in one batch, because many small plants have operating staff present during only one shift. With adequate capacity, these tanks can successfully be used to gravity thicken the biosolids following stabilization. When this is done, special equipment must be used to withdraw the thickened biosolids.

In continuous mixing facilities, the pH and volume are held constant. This type of system requires automatic lime feeding equipment. The primary advantage is that a smaller tank may be used than that used for batch mixing. Because pH is one of the most important factors, the tank contents should be monitored closely and maintained at a pH of more than 12 for at least 2 hours following mixing.

Mixing systems must provide adequate mixing to keep the solids in suspension and to distribute the lime efficiently. The two most common mixing systems are diffused air and mechanical types. Both have been successful; however, diffused air is more widely used.

Diffused air mixing systems have at least two important advantages over mechanical systems. One is increased aeration, which helps keep the feed fresh preceding lime addition in batch-type operations. The second advantage is that there is less potential for debris to foul the equipment (however, "nonclog" mechanical systems are available).

Diffused air systems also have several disadvantages. One is that ammonia stripping creates odors and reduces the fertilizer value of the product. Ammonia release can also be hazardous, and adequate ventilation must be provided. A second disadvantage is that the mixture absorbs carbon dioxide from the air; therefore, more lime is required to raise pH because part of the lime added reacts with the carbon dioxide. Finally, because gases such as ammonia are stripped, the facilities must be enclosed and the offgas may require treatment.

Design criteria for mixing facilities are similar to those for aerobic digestion facilities. When a diffused air system is used for mixing, the diffusers should be of the coarse bubble type. Diffusers typically are mounted along one wall of the tank to induce a spiral roll mixing pattern. Air flow rates for mixing of 0.3 to 0.5 $L/m^3 \cdot s$ have been used successfully (Beals, 1976). Air flow requirements may be higher for mixing thickened feeds.

Mechanical mixing design criteria are based on the use of bulk fluid velocity and impeller Reynolds number. Table 6.8 lists the various sizes of mechanical mixers required for various volumes. This table is based on maintaining the bulk fluid velocity (turbine agitator pumping capacity divided by cross-sectional area of mixing vessel) greater than 0.13 m/s and an impeller Reynolds number greater than 1 000. The values in Table 6.8 are adequate for mixing feeds with concentrations of up to 10% dry solids and viscosities up to 1 Pa/s (1 000 cP).

When the feed is conditioned in mixing tanks before thickening or dewatering, careful consideration to mixing design is required to prevent floc shearing. In general, lower mechanical mixer speeds and larger turbine diameters are required. A variable-speed drive is desirable for mechanical mixing to allow process control. For additional conditioning requirements, consult *Sludge Conditioning* (WPCF, 1988).

Several additional publications deal with selecting lime and lime-handling equipment and designing lime application systems (AWWA, 1983, and NLA, 1988). These should be consulted for additional design information.

DEWATERED CAKE/CHEMICAL MIXING FOR DRY ALKALINE STABILIZATION. Mixing or blending dewatered cake and alkaline material is the most critical component of dry alkaline stabilization. The goal is to provide intimate contact between the cake and chemical, resulting in a com-

Table 6.8   Mechanical mixer specifications for liquid lime stabilization (Counts and Shuckrow, 1975).[a]

Tank size		Tank diam		Motor size		Shaft speed, rpm	Turbine diam	
m³	gal	m	ft	kW	hp		m	ft
19	5 000	2.9	9.5	6	7.5	125	0.8	2.7
				4	5	84	1.0	3.2
				2	3	56	1.1	3.6
57	15 500	4.2	13.7	15	20	100	1.1	3.7
				11	15	68	1.3	4.4
				7	10	45	1.6	5.3
				6	7.5	37	1.7	5.6
114	30 000	5.2	17.2	30	40	84	1.5	4.8
				22	30	68	1.6	5.1
				19	25	56	1.7	5.5
				15	20	37	2.1	6.8
284	75 000	7.1	23.4	75	100	100	1.6	5.2
				56	75	68	1.9	6.2
				45	6	56	2.0	6.6
				37	50	45	2.2	7.3
380	100 000	7.8	25.7	93	125	84	1.8	6.0
				75	100	68	2.0	6.5
				56	75	45	2.4	7.8

[a] Bulk fluid velocity >0.13 m/s (26 ft/min).
Impeller Reynolds >1 000.
Mix tank configuration:
   Liquid depth equals tank diam.
   Baffles with a width of 1/12 the tank diam, placed at 90-deg spacing.

plete pH adjustment throughout the cake/chemical mixture. The lack of adequate cake/chemical mixing has led to incomplete stabilization, odors, and dust problems at several U.S. dry alkaline stabilization facilities.

The mixing step typically is accomplished using a mechanical mixer, such as a pug mill or plow blender. Typical cake/lime pug mill and plow blender mixers are shown in Figures 6.12 and 6.13, respectively. Dewatered cake and chemical are added together at the "head" of the mixer. Both batch and continuous mixers are available. The proportioning of the alkaline material dosage rate to the rate at which cake is introduced is important.

Cake/chemical mixing is an inexact science. A large number of variables will affect the mixing process and, consequently, the characteristics of the biosolids product. Mixer selection typically is based on past experience and trial-and-error testing. Many mixer manufacturers have mobile pilot full-scale testing equipment available. Whenever possible, this equipment should be used in the evaluation and selection of the proper and most effective mixing equipment.

The mixing characteristics of a dewatered cake will vary with solids concentration, polymer used for conditioning before dewatering, chemical type and dosage, temperature, mixing intensity, available mixer equipment surface area per volume exposed to the cake, and mixing retention time. When selecting a mixer, consideration should also be given to minimum and maximum cake production, hours of operation, and other operating conditions. To

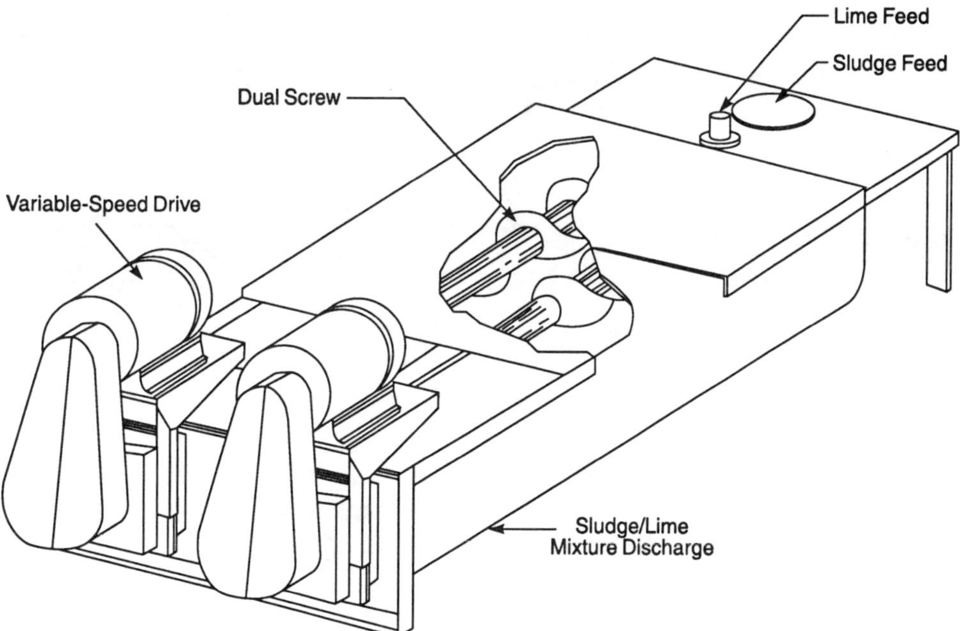

Figure 6.12   Typical dual screw pug mill mixer.

account for these variable mixing characteristics and conditions, mixers can be equipped with variable-speed drives, adjustable mixer paddle configurations, weir plates, and other options to adjust mixing intensity and retention time (Christy, 1992).

Physical characteristics of the alkaline product will vary with the mixing parameters. Biosolids product physical consistency can range from sticky and plastic to granular and dusty. The goal of the mixing step is to produce a product compatible with the end-use or next processing step. Characteristics of the alkaline biosolids product may continue to change for a period of time up to several days after the mixing step because of ongoing chemical reactions, temperature, and other parameters.

SPACE REQUIREMENTS. Depending on the site constraints, the type of process used, and the amount of solids to be processed, site preparation for alkaline stabilization processes typically is minimal. The process equipment typically can be arranged to accommodate various site constraints. Because processes are relatively simple to operate and do not require extensive, complex alkaline stabilization equipment, they can be implemented in a short time frame and in a relatively small space. Mobile, skid-mounted equipment can be used for backup or in a emergencies, as well as in demonstration programs to encourage interest in the product. The layout used for one 91-Mg/d (100-ton/d) advanced alkaline stabilization facility is shown in Figure 6.14.

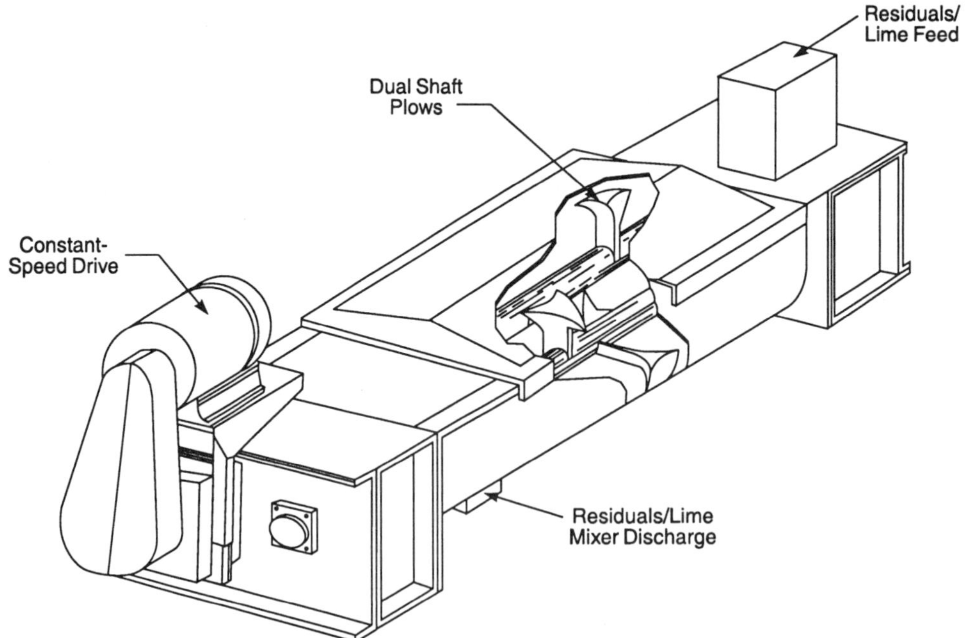

Figure 6.13   Typical plow blender mixer.

Space requirements for an alkaline stabilization facility include

- Solids processing area,
- Drying area (if necessary), and
- Product storage area.

Land-area requirements will vary depending on the process to be used; the amount, type, and characteristics of solids to be processed; and the specific site. The area requirements for drying and curing are also highly site-specific and depend on the amount of material to be dried/cured and the drying method used. Typical area requirements for drying/curing are 25 to 34 m^2/Mg (300 to 400 sq ft/wet ton) of processed cake. The drying/curing building size can be significantly reduced by increasing the alkaline chemical dose rate or using a mechanical dryer.

Consideration should be given to storing product off site if not enough area is available at the processing site. Landfills typically can provide adequate area to accommodate the drying/curing process. However, the drying and storage areas must be relocated as the landfilling progresses. The drying area must also be easily accessible and large enough for trucks to unload the product without excessive maneuvering. Land area will also be required to accommodate additional truck traffic on site. Outdoor drying/curing at landfills can cause odor complaints.

If the alkaline stabilization process is located at the WWTP, the access roads most likely already exist. Sufficient access should be provided for

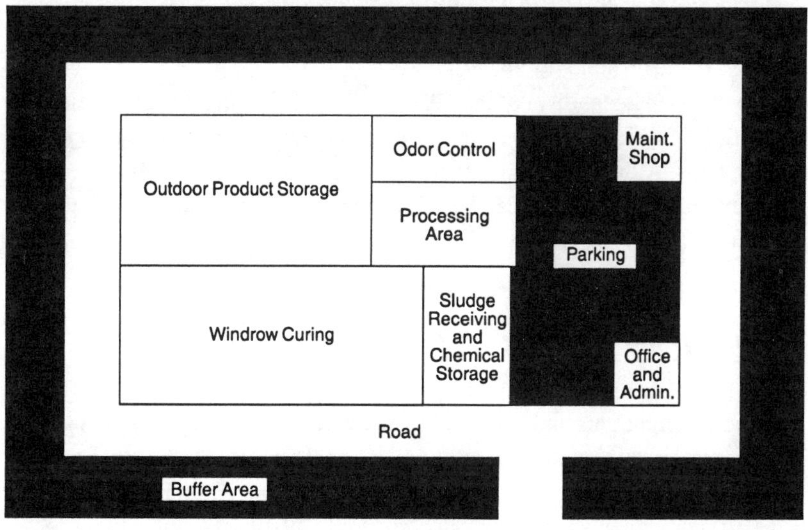

Figure 6.14   Layout for a 91-Mg/d (100-ton/d) advanced alkaline stabilization facility (Fergen, 1991).

regular delivery of alkaline materials. This truck access should not interfere with the traffic associated with the process or with product distribution.

ECONOMIC CONSIDERATIONS. Economics is another important factor when selecting a solids management option. The costs of an alkaline stabilization process should be evaluated with respect to the total cost of the option over its useful life using present worth, equivalent annual cost, or similar measures. The cost of disposal or beneficial use of the product can be significant—$11 to $165/dry Mg ($10 to $150/dry ton)—for hauling and land application and must be included in the cost analysis (Jacobs *et al.,* 1992). In addition, the costs of a privatized option, if that is the preferred procurement method, should be compared with alternative solids management options that are to be owned and operated by the municipality.

Annual operation and maintenance costs include operating labor, chemical costs, transportation of the chemicals to the facility and the product to sites, fuel, utilities, and maintenance costs. Other annual costs may include public education and relations; marketing; biosolids, soil, and agronomic testing; and analyses. An owner should exercise caution in examining the annual costs at other facilities. The cost includes a number of site-specific factors such as the local labor cost, distance of the chemical supply from the facility, and power. Moreover, minimum biosolids production amounts specified in the contract also affect the total annual costs. When alkaline stabilization technologies are operated under private contracts, it is important that a contract be negotiated that accurately reflects the actual biosolids production.

Many site-specific factors influence costs and make economic evaluations and comparisons difficult. The site-specific factors include physical layout, feed type and characteristics, product use, and local regulations. Additionally, at some alkaline stabilization facilities, existing equipment has been retrofitted and used in the alkaline stabilization process.

Flexibility (adaptability) of the alkaline stabilization process and the use of existing facilities should be considered when evaluating solids management options. Although it may not always be possible to use or retrofit existing equipment, cost savings can be achieved by the municipality if existing equipment is used as part of the process train.

*O*PERATIONAL CONSIDERATIONS

This section highlights and summarizes some of the operation and maintenance issues pertaining to alkaline stabilization. In general, most alkaline stabilization technologies are relatively simple and are not equipment-intensive, and staff requirements are low when compared with other stabilization processes. Operational considerations that need to be addressed include start-up issues, labor requirements, health and safety considerations, feed and product quality monitoring, maintenance, odors, dust, drying, procurement options, and process performance.

START-UP ISSUES. Start-up issues associated with alkaline stabilization will vary depending on the installation. Issues of greatest concern include equipment performance; process verification; physical, chemical, and biological product quality to verify regulatory compliance and ensure product acceptance; and, if privatized, contractor performance.

While the alkaline stabilization processes are not as equipment-intensive as other stabilization processes, some equipment problems and operational difficulties may occur during start-up. In efforts to optimize the dry chemical dose and produce the desired product, it may be necessary to vary the mixer paddle speed and retention times. All process equipment should be tested at rated capacity during the start-up period. A significant amount of representative feed should be tested and accurate dosage measurement and homogeneous mixing should be verified. If several types of alkaline materials are to be used, each should be tested with the system storage and feeding equipment to verify acceptable operation.

Regulatory officials should be consulted before start-up to verify the parameters to be monitored for approval of the process. The results of this monitoring should be submitted to the officials as soon as possible to initiate the approval process. Delays because of the regulatory/permits process are not uncommon. Appropriate measures should be taken to avoid these delays if at all possible. Frequent and continued communication with key regulatory personnel can facilitate the approval process. Appropriate notification must also be made to the federal permitting authority (U.S. EPA region or delegated state agency) before start-up.

The start-up operations provide an opportunity to vary process parameters and evaluate the effects of these changes on product quality. Although the effect of various chemical dosages and drying times should have been evaluated during pilot- or bench-scale testing, pilot-scale test conditions do not always adequately simulate full-scale operation.

An advantage of these processes is the ability to start up operations in a short timeframe. A mobile, outdoor processing unit can be fully operational in approximately 10 days or fewer depending on the amount of material to be processed.

LABOR REQUIREMENTS. Work activities during normal operation include process and equipment monitoring, routine inspection and maintenance, and general utility work. The number of operators required to perform these activities depends on the facility size, the layout of the WWTP, and the number of operating shifts.

Because alkaline stabilization processes are relatively simple and not equipment-intensive, staffing requirements are minimal compared with other stabilization processes. Typically, one operator is responsible for monitoring the control panel and the feed and chemical conveyance to the mixer. Another operator takes samples and monitors the pH, and a third operator is available on a rotation basis and assists in routine housekeeping, maintenance, and operation. Other operators may be needed for solids drying/curing in windrows and product loading, hauling, and application. Extensive operator training is not required for these alkaline stabilization processes. Because most of the equipment is common to many WWTPs (belt conveyors, pug mill mixers), many operators have the necessary experience and equipment skills to run the process. However, operators should be educated in the proper use, maintenance, and calibration of the monitoring and control equipment.

The working relationship between the WWTP operators and the privatized system personnel, if used, is important. The roles and responsibilities of all parties should be clearly defined to reduce the potential for conflicts, especially in WWTPs with unionized work forces.

HEALTH AND SAFETY CONSIDERATIONS. With few exceptions, health and safety considerations for alkaline stabilization are no different than for typical WWTP operations. Standard Occupational Safety and Health Administration (OSHA) requirements are maintained and include the use of safety glasses and hard hats.

The dust generated from the alkaline materials is likely to be the most significant health concern. Alkaline materials are caustic and cause skin burns and irritation and discomfort to moist surfaces (such as eyes, lips, and sweating arms); therefore, readily accessible eye washes and showers should be provided at various locations throughout the WWTP. The operators working in dusty environments or servicing alkaline storage and feeding equipment should be supplied with proper work clothing and safety equipment including gloves, proper respirators, and eye protection.

Ammonia is another safety concern, especially if anaerobically digested biosolids are processed, because it is likely that considerable ammonia gas will be released (approximately 6 to 10 times when compared with raw solids). Ammonia emissions can be controlled through proper ventilation of mixers, storage hoppers, and loadout areas. Strong releases of ammonia may be experienced during mechanical aeration/mixing and drying. Process mixing equipment should be enclosed and vented to odor-control facilities if at all possible. It may be necessary, in some areas, to provide operators with respirators, depending on the amount of ammonia released to meet OSHA requirements.

Special safety measures may be required for drying areas. The layer of fine dust that may tend to settle from operations at the drying area has been found to be slippery on concrete or asphalt surfaces. During wet weather, a layer of mud may form outdoors on the drying pad. Mud is slippery and may pose a hazard to pedestrians and vehicle traffic. Special precautions should be taken to improve safety through good housekeeping practices.

PROCESS MONITORING AND CONTROL. It is critical that the feed and alkaline materials be monitored frequently so that process adjustments can be made to achieve adequate stabilization and a consistent product. The effects of incomplete stabilization are not readily apparent and may not be seen at the WWTP; therefore, proper process control is very important. Operators must be aware that absence of odors at the WWTP and acceptable dewatering characteristics alone are not good indicators of adequate stabilization.

Monitored characteristics include total solids, pH, and temperature of both the feed and the product. For Class A (PFRP) products, fecal streptococcus indicator organisms must be monitored at the frequency specified in 40 CFR Part 503.16 (U.S. EPA, 1993). Metals must be monitored if the product is to be used for agricultural purposes. Toxic characteristics leaching procedure tests must be performed if the product will be landfilled. Quality control data may be required for regulatory approval. The method and frequency depended on the regulatory requirements. In some cases, odor characterization and emission monitoring may also be required.

Chemical dosage adjustments may be made in response to manual measurements of temperature and pH, as well as visual inspection of the feed/chemical mixture. However, some automatic process control can be incorporated if desired. Weighing and volumetric systems should be calibrated monthly. For instance, thermocouples can be used to measure the heat pulse if an enclosed vessel is used. Chemical feed rate could be controlled by pacing it with the incoming feed flow rate or dewatered cake using a weigh belt or similar means.

Programmable logic control of the chemical feed system aids in producing a consistent product. A typical system may include an electronic chemical metering system linked to feed weight belt scales. Chemical feed is automatically controlled based on the weight of the feed solids. With this type of system, the reagents are added in proportion to the amount of dewatered cake. Special care should be taken to keep the weigh scales calibrated frequently and operating correctly to ensure that the appropriate dosage of alkaline materials will be added. An automated system also decreases the number of personnel needed to operate the process.

Sensors, particularly pH electrodes (both laboratory and automatic process control units), must be properly cleaned, calibrated, and maintained. Special pH electrodes are necessary for routine measurements of more than pH 10. The pH must be carefully monitored to ensure it is maintained high enough for a sufficient period. Portable pH pen probes are acceptable for process monitoring. Microbiological examination for indicator organisms (fecal coliforms, fecal streptococci) should be performed regularly by a qualified laboratory.

MAINTENANCE. The successful operation of an alkaline stabilization process depends on the reliability and performance of equipment. Routine inspection and servicing should be important elements in a preventive maintenance program. Motor fans, conveyor bearings and seals, and screw flights should be inspected regularly and cleaned or lubricated as necessary. Daily and weekly inspection schedules should be followed to prevent excessive wear of equipment and downtime as a result of improper maintenance and the frequent buildup of scale on equipment from alkaline materials.

An adequate supply of spare parts is critical to the operation of any process equipment. A spare parts inventory should be developed, taking into consideration those parts that are prone to frequent failure and their importance to the process. Parts that are essential to continued operation should be kept in stock.

A redundant equipment train should be strongly considered during the project development. A second train of equipment would allow routine maintenance and calibration without downtime and would provide operational flexibility.

ODOR GENERATION AND CONTROL. Odors and odor control are important issues when evaluating alkaline stabilization as a solids management option. Inadequate control and treatment of odors can be detrimental to a solids management program. Local conditions such as weather, other odor sources, and the characteristics of the odor-causing compounds will influence the selection and design of an odor treatment system. There are many site-specific factors that should be considered when developing a publicly acceptable odor-control program.

Elements of a successful odor-control effort include

- Initial site selection,
- Proper process performance,
- Reducing product storage time and volume,
- Identification of odor sources and odor-causing compounds,
- Meteorological modeling at different heights,
- Distance to nearest receptors, and
- Appropriate odor-control technology and equipment.

Ammonia odors are most typically encountered at alkaline stabilization facilities. The elevated pH resulting from the addition of alkaline materials causes the dissolved ammonia in the liquid sludge to be released as a gas. Although the odors tend to dissipate quickly, the ammonia levels in mixing and drying areas can be high if the gas is not collected and treated. As a result, if adequate ventilation is not provided, operators may need to wear respirators. Appropriate odor-control technology and equipment should be provided at alkaline stabilization facilities to ventilate and scrub the air, to remove ammonia, to reduce odor problems, and increase public acceptability. The intensity of the ammonia emissions in the processing area may mask other odors that are more prevalent as pH and temperatures are elevated and that do not readily dissipate, such as trimethylamines. Consequently, an odor survey should be performed to identify the sources of odors and to characterize the odor-causing compounds.

In addition to an odor survey, an assessment of meteorological conditions and atmospheric dispersion should be performed. Atmospheric data should be collected on wind speed and direction, temperature, and inversion conditions. This information typically is available from local weather stations and can be used to determine the effect of the odor on residents near the alkaline stabilization facility and the degree of odor control necessary to meet community odor standards.

An effective odor-control program involves operational monitoring and may include bench-scale testing (to determine ammonia emissions at various chemical dosages) and gas chromatography/mass spectrometry testing. After the odors have been characterized, the next step is to collect and treat the odors. Pilot-scale testing is also helpful for checking the effectiveness of a proposed treatment option and its chemistry. At many alkaline stabilization facilities, odor control consists primarily of dilution of odors through open-air drying. If drying operations are enclosed, dilution and dispersion through rooftop ventilation can be used. However, if large quantities of materials are processed in a densely populated area, a combination of dispersion and chemical scrubbing should be seriously investigated. It is important that the odor-control program be responsive to odor complaints. Taking into consideration the meteorological conditions and the sources and types of odors, operation or process modifications may be necessary to resolve the problem.

It was initially thought that odor-control systems were not necessary on alkaline and advanced alkaline stabilization facilities. However, numerous odor concerns and complaints have resulted in a recent understanding that odor-control systems should be strongly considered and may be necessary for alkaline stabilization systems located near populated areas. Odor-control systems for alkaline stabilization systems may be enhanced ventilating systems and simple one-stage chemical scrubbers designed for ammonia removal only in the feed/chemical mixing area. Odor-control systems can also be state-of-the-art, three-stage, packed tower/mist scrubber/packed tower odor-control systems with air dispersion stacks design to treat high volumes of foul air containing particulates, ammonia, amines, dimethyldisulfide, mercaptans, and hydrogen sulfide generated from all areas of the solids processing and alkaline stabilization process equipment. These sophisticated odor-control systems use sulfuric acid and sodium hypochlorite and sodium hydroxide odor-control chemicals to neutralize and oxidize the odor-causing compounds.

DUST. Dust is a persistent problem inherent to alkaline stabilization systems. Alkaline material handling systems have the potential to create significant dust problems, particularly if fine-textured materials such as hydrated lime, CKD, or LKD are used. Because of the caustic nature of alkaline materials, operators working in high dust areas should wear gloves, goggles, and dust masks. Design of the alkaline materials handling system should include provisions to reduce dust production. Excessive alkaline dust will also affect odor-control scrubber performance, including acid chemical requirements.

Producing a material with soillike characteristics can also cause dust concerns. If the process includes a drying step or the addition of chemicals in large doses, significant quantities of dust may be generated. When an alkaline-stabilized product is overdried to more than 65 to 70% solids concentration, transfer of the product in any manner, such as with front-end loaders, can generate dust. Good site housekeeping measures, such as frequent washdown and street sweeping, are recommended (Sloan, 1992).

SIDESTREAM EFFECTS. Alkaline stabilization processes typically will have little effect on WWTP operation. Minimal sidestreams may result from site drainage and product stockpile leachate and runoff if the storage area is not covered. However, a potential WWTP sidestream load is ammonia recycle if ventilation and odor-control acid scrubbing systems are installed.

DRYING. Supplemental drying, if required by the process, also requires special consideration. Drying may make the product easier to handle, and the type of drying system will affect the characteristics of the product. For instance, if the material is set out on a pad for solar drying without mechanical turning, it may dry in large clumps that would be incompatible with the final use if the product was to be land-applied with granular fertilizer spreaders.

The duration of the drying/curing process is affected by environmental conditions, chemical dose, windrow configuration, and initial and final solids concentrations. The duration of curing depends on the time required to achieve the process goal (that is, for the heat of reaction to occur and for the pH to rise high enough to achieve pathogen destruction). Drying is simply the removal of moisture. Objectives of both drying and curing processes are to modify physical properties to attain the desired product solids concentration and characteristics and to reduce the heat loss during curing so that the chemical dosage is effective. The duration of drying to achieve a desired solids concentration in the product is affected by the size of the windrow and weather conditions, if the drying facility is not enclosed.

PROCESS PERFORMANCE. Properly designed and operated alkaline stabilization systems reduce odors, odor production potential, and pathogen levels. The nature and extent of these effects are addressed in the following paragraphs.

Odor Reduction. With proper mixing, alkaline stabilization systems substantially reduce odor. One of the odor sources in solids-processing facilities, hydrogen sulfide, is essentially eliminated after the alkaline chemical is added and the pH increases to 9 or above, because it is converted to nonvolatile ionized forms (see Figure 6.15). Initially, when air-mixing systems are used, ammonia odors increase as a result of ammonia stripping. After the initial ammonia odors have been emitted and dispersed or treated, odors can be reduced by a factor of ten (Westphal and Christensen, 1983). Other odorous gases emitted at high pH and temperature such as trimethylamine must be considered and dispersed or treated at alkaline stabilization facilities.

Pathogen Reduction. Several studies have demonstrated that both liquid lime and dry lime stabilization achieve significant pathogen reduction, provided that a sufficiently high pH or temperature is maintained for an adequate period of time (Bitton *et al.*, 1980, and Christensen, 1982). Table 6.9 lists bacteria levels measured during the full-scale studies at the Lebanon, Ohio, WWTP and shows that liquid lime stabilization at a pH of 12.5 and a 25% dry-weight dose reduced total coliform, fecal coliform, and fecal streptococci concentrations by more than 99.9%. The numbers of *Salmonella* and *Pseudomonas aeruginosa* were reduced below the level of detection. Table 6.9 also shows that pathogen concentrations in liquid lime-stabilized biosolids ranged from 10 to 1 000 times less than those in anaerobically digested biosolids from the same WWTP.

Christensen (1987) researched the pathogen reduction performance of dry lime stabilization using dry quicklime doses of 13 and 40% on a dry-weight

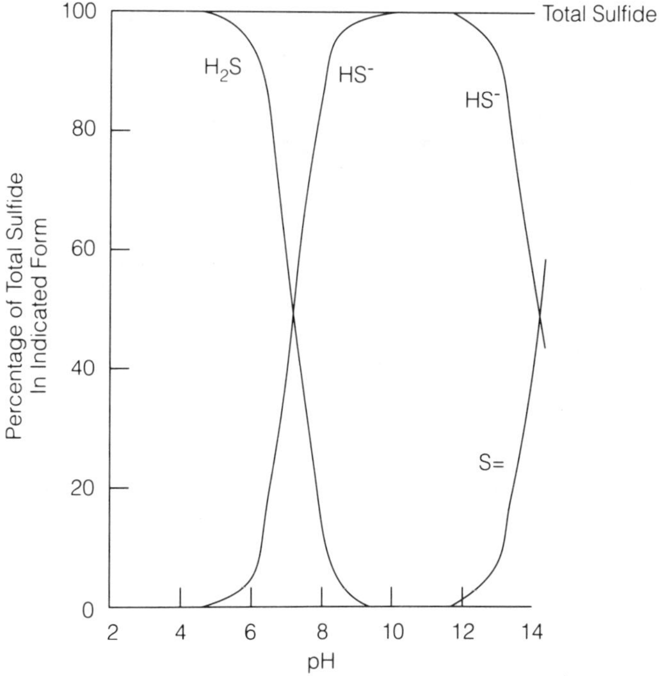

Figure 6.15   Effect of pH on speciation of hydrogen sulfide (WEF, 1992).

basis as calcium hydroxide. His results indicated dry lime stabilization can reduce fecal coliform and streptococcus pathogens by at least two orders of magnitude. This was as good as, and in some cases better than, the results of standard liquid lime stabilization and liquid lime conditioning followed by vacuum filtration (Figures 6.16 and 6.17). No growth of either fecal coliform or fecal streptococcus occurred by day 7 (Westphal and Christensen, 1983). Westphal and Christensen (1983) also reported that alkaline stabilization processes performed as well as or better than mesophilic aerobic digestion, anaerobic digestion, and mesophilic composting in reducing the densities of fecal coliform and fecal streptococcus (Table 6.10). Additional discussions of lime treatment and the control of bacterial, viral, and parasitic pathogens is reviewed in reports of Christensen (1987) and Reimers *et al.* (1981).

Additional studies on dry lime stabilization showed 4 to 6 log reduction of fecal streptococcus at a pH of approximately 12 (Otoski, 1981). Such a treatment scheme can yield Class A biosolids by using a 2-to-1 lime dose at 20% solids on a dry weight basis to raise the temperature of the solids to more than 70°C for 30 minutes and meet pathogen reduction requirements as a result of the heat of lime hydration.

Data indicate treatment of dewatered municipal wastewater cake with CKD, alone or with a small amount of quicklime with CKD, will reduce

Table 6.9 **Reduction in bacteria by liquid lime stabilization at Lebanon, Ohio (U.S. EPA, 1979).**

Type of solids	Bacterial density, number/100 mL				
	Total coliforms[a]	Fecal coliforms[a]	Fecal streptococci	*Salmonella*[b]	*Ps. aeruginosa*
Raw sludge					
Primary	2.9×10^9	8.2×10^8	3.9×10^7	62	195
Waste activated	8.3×10^8	2.7×10^7	1.0×10^7	6	5.5×10^3
Anaerobically digested biosolids					
Mixed primary and waste activated	2.8×10^7	1.5×10^6	2.7×10^5	6	42
Lime-stabilized biosolids[c]					
Primary	1.2×10^5	5.9×10^3	1.6×10^4	<3	<3
Waste activated	2.2×10^5	1.6×10^4	6.8×10^3	<3	13
Anaerobically digested	18	18	8.6×10^3	<3	<3

[a] Millipore filter technique used for waste-activated sludge. Most probable number technique used for other sludges.

[b] Detection limit = 3.

[c] To pH equal to or greater than 12.0.

pathogenic microbial populations to below U.S. EPA's Class A standard (Burnham *et al.*, 1992). Both laboratory- and large-scale field tests have shown that indigenous and seeded populations of salmonella, poliovirus, and ascaris ova could be eliminated within 24 hours if the treated biosolids are contained at pH of 12 and 52°C for 12 hours.

There is little information about the amount of virus reduction during lime stabilization, but lime has been identified as an effective viricide. Qualitative analysis has indicated substantial survival of higher organisms (such as hookworms and amoebic cysts) after 24 hours at high pH (Farrell *et al.*, 1974). It is not known whether prolonged contact would eventually destroy these organisms. Class A alkaline stabilization processes that maintain 70°C for 30 minutes have been shown to kill *Ascaris* ova.

PROCUREMENT OPTIONS. Many advanced alkaline stabilization technologies are offered by private firms and require the use of proprietary processes or specialized equipment. Because of the proprietary nature of the processes, royalty fees, quality control fees, or sole-source equipment may be involved. Additionally, some of the firms may offer turnkey design–build

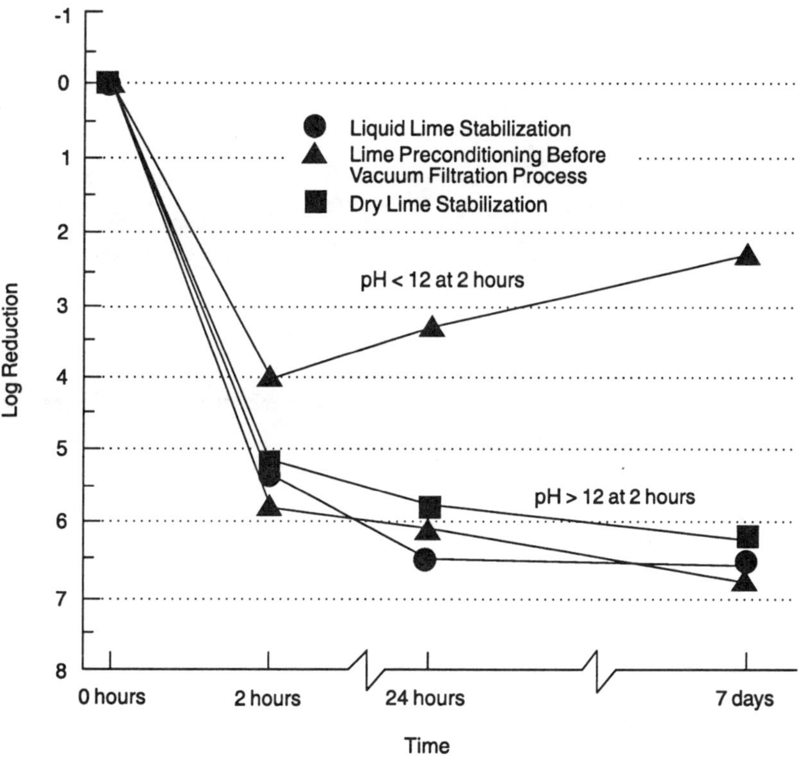

Figure 6.16   **Average fecal coliform inactivation by two liquid lime stabilization processes and one dry lime stabilization process (Westphal and Christensen, 1983).**

facility procurement options or require privatization of various types of solids processing services.

A privatized facility could be designed, constructed, equipped, commissioned, and operated as a commercial enterprise, all by a private firm. The sludge would be accepted from the municipality as a customer, and a fee charged for the stabilization service. Depending on the degree of privatization preferred by the municipality, the commercial firm assumes most, if not all, of the responsibility and liabilities. The commercial firm can be responsible for the entire process from permitting and pilot testing to operation and marketing of the product. These services are reflected in the total cost of the process. Care should be exercised when negotiating the contracts for private services to ensure the municipality is not penalized for poor performance by the contractor and that the contractor receives a fair return for the services (Stone *et al.*, 1991).

There are several advantages to a privatized approach. First, a smaller commitment of the municipality's staff is necessary. The majority of time committed to the project is spent on the negotiation of a service agreement. In some cases, negotiation of a service agreement can be a lengthy process. Also, the

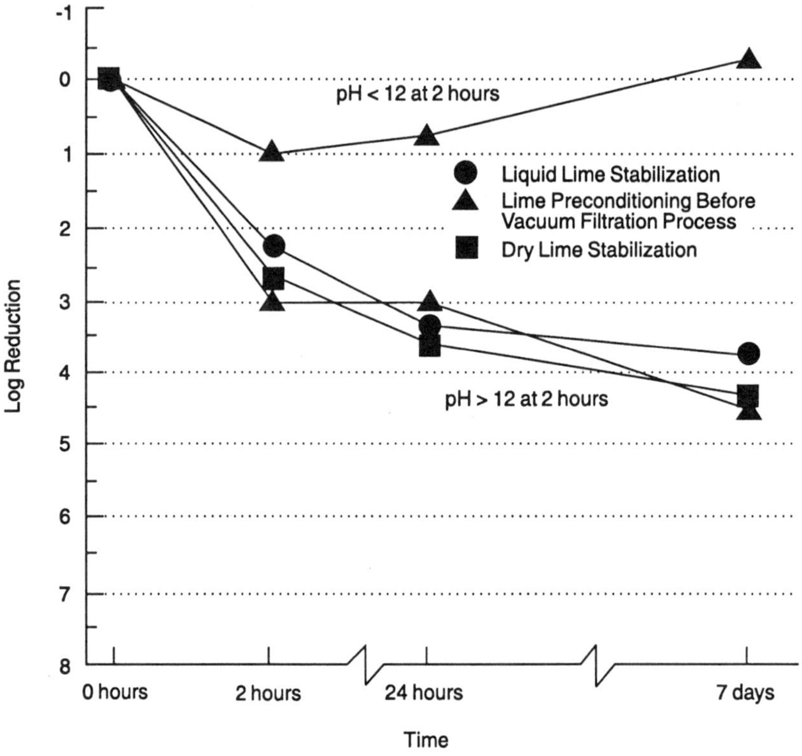

Figure 6.17 **Average fecal streptococcus inactivation by two liquid lime stabilization processes and one dry lime stabilization process (Westphal and Christensen, 1983).**

privatized alkaline stabilization facility can be implemented in a relatively short timeframe, as compared with conventional procurement of traditional solids management facilities.

Privatization also has several disadvantages. First, the municipality may have little control over the type of facility constructed. This may not be a primary concern because the municipality is not investing any capital. However, the municipality must pay a service fee, which reflects amortized capital and expected operating costs. Original fees may be negotiated before the construction of the facility but typically are revised annually or biennially. The municipality typically must guarantee a minimum feed quantity and quality and must pay for the treatment of the negotiated amount, regardless of the actual amount treated.

Another concern is public relations. The public may blame any nuisance problems or permit violations on the municipality, which may have no authority or responsibility to solve the problem. Regulatory agencies are debating the issue of responsibility in these cases. Generators have not been guaranteed a clear release from responsibility for improper solids management or excessive odor generation by private facilities. Some of this responsibility can be contractually transferred to the private firm, but typically at a high price. The privatized facility is designed, operated, constructed, equipped, and

Table 6.10   Reduction of bacteria by different stabilization processes.

Process	Fecal coliform	Fecal streptococcus
Anaerobic digestion (35°C)		
Mean	1.84	1.48
Range	1.44 – 2.33	1.1 – 1.94
Aerobic digestion		
20°C[a]	1	1
30°C[a]	2	1.64
Composting	≥4	2.9
Liquid lime stabilization		
Raw primary	5.1	2.4
Waste activated	3.2	3.2
Mixed primary and trickling filter humus, 4% solids	2.6	1.8
Storage[b]		
10°C	—	1
20°C	—	1.5
30°C	—	2.0

[a] Laboratory study, 35-day detention time.
[b] Laboratory study, 30-day detention time.

eventually operated as a commercial enterprise by a private firm. The costs of a privatized option, if that is the preferred procurement method, should be compared to alternative management options that are to be owned and operated by the municipality.

PRODUCT END-USE CONSIDERATIONS

Alkaline-stabilized technologies can result in several types of beneficial products. Alkaline materials can be custom-blended with the feed to produce a specific end-use market product. Each end-use product is subject to particular quality requirements and standards. Potential options for alkaline-stabilized products include the following:

- Agricultural (organic fertilizer, ag-lime substitute, or soil amendment),
- Nonagricultural land application/land reclamation/dedicated land disposal,
- Landfill (disposal or daily, intermediate, final, and vegetative cover),
- Manufactured organic topsoil blends,
- Bulk fill applications (slope stabilization, dike construction), and
- Horticulture (nurseries, sod farms).

The material can be applied to agricultural land as a fertilizer, soil conditioner, or liming agent. Wastewater biosolids contain organic matter and plant nutrients, making them a valuable crop fertilizer and soil conditioner. However, the addition of alkaline material results in the dilution of some plant nutrients. Additionally, ammonia nitrogen in the biosolids is volatilized during the alkaline treatment process. Tests on alkaline-stabilized biosolids have also shown that a substantial portion of the organic nitrogen initially mineralized is adsorbed, which further reduces the available nitrogen amount to the plant (Logan, 1990). The net result may be a relatively low-grade fertilizer, but a good lime substitute and organic soil amendment. Because its plant nutrient content may be low, the application rate of alkaline-stabilized biosolids for agronomic purposes may be limited based on calcium carbonate equivalence or alkalinity content of the biosolids product.

An advantage of an alkaline-stabilized biosolids product over a traditional biosolids product (such as compost) is that it can partially or fully satisfy the liming requirements for many soils. Another important factor is that the alkaline material may contain small amounts of plant nutrients, thereby enhancing the fertilizer value of the product. For instance, most CKD contains significant amounts of potassium, as well as smaller amounts of trace nutrients. However, some CKD has been reported to have significant lead concentration (570 ppm) (Christy, 1992).

Another end use for alkaline-stabilized biosolids is daily, intermediate, final, and vegetative cover material for landfills. However, most regulatory authorities require extensive testing and documentation before they allow the use of an alkaline-stabilized biosolids product as landfill cover material. Demonstration programs typically are the first step in obtaining regulatory approval. Important considerations for landfill use are product odor potential, percent solids concentrations, nonhazardous determination, bearing strength, ignitability, compaction, permeability, and compatibility with materials handling and compaction equipment. Since 1989, the city of Oxnard, California, has blended biosolids with gypsum fly ash to successfully produce daily cover for the municipal landfill. This project has saved both the wastewater and solid waste divisions of the city money by eliminating the need to import soil cover material.

Other potential uses of alkaline-stabilized biosolids are currently being investigated. These include use for land reclamation projects, structural fill material, turf establishment, topsoil manufacture, and soil farms.

Land reclamation is a promising future use of alkaline-stabilized biosolids products. Other biosolids products, such as compost, have been used successfully for land reclamation at a number of locations. An advantage of alkaline-stabilized biosolids over other biosolids products is the ability to raise the pH of the predominantly acid soils at most reclamation sites.

Alkaline-stabilized biosolids have been used as a structural fill material in dike construction in Wilmington, Delaware. In Waukegan, Illinois, fly-ash-stabilized biosolids have been used as a structural material in a monofill (Byers and Jensen, 1990). Other uses as structural material are possible, due in part to the pozzolanic properties of the alkaline stabilization materials.

The potential for achieving a higher degree of stabilization (such as Class A) offers a number of potential end uses through product distribution and marketing. In general terms, distribution and marketing refers to the unrestricted programs that sell or give away biosolids products. Traditional distribution and marketing programs have included the use of compost or thermally dried pellets for landscaping and horticultural applications. It is important to realize that composted and alkaline-stabilized biosolids products differ physically and chemically and should be used in different ways. Alkaline-stabilized products typically have characteristics (such as high pH and salt content) that significantly reduce their effectiveness for horticultural uses (*Municipal Sewage Sludge,* 1992). Therefore, alkaline-stabilized products typically cannot compete with compost products in the horticultural markets.

Establishment and maintenance of turf is a potential use for alkaline-stabilized biosolids. The soillike characteristics and near absence of odor of alkaline-stabilized biosolids should make them suitable for turf applications. Equipment designed for use with granular fertilizer may be used to apply the dried biosolids product. Low odor emission potential is a necessity when considering the extent of public exposure associated with most turf applications.

Production of manufactured organic topsoil is a nearly untapped use for alkaline-stabilized biosolids. Composted biosolids have been used extensively in this area. Alkaline-stabilized biosolids can make a good replacement for topsoil if they are blended with other materials such as peat moss, sand, yard waste compost, or subsoil. Markets for this application are likely to develop near large metropolitan areas where good topsoil is scarce and expensive. The biosolids product must have soillike characteristics to be blended acceptably with the other materials. Use of alkaline-stabilized biosolids alone as potting soil, however, is somewhat limited, because of their high pH and salt content (*Municipal Sewage Sludge,* 1992).

The quality of the product must be compatible with the intended use. Characteristics such as stability, pH, liming value, nutrient content, percent solids, granularity, and permeability should be monitored and controlled for specific end-use markets.

The quality of the product also affects the allowable application rate. Depending on the concentration of metals, the application rate of the alkaline-stabilized biosolids product may be limited on either an annual or cumulative basis. If the product is used for agronomic purposes, the allowable application rate will generally depend on the nutrient or liming value of the product (Rubin, 1991).

*R*EFERENCES

American Water Works Association (1983) *Standard for Quicklime and Hydrated Lime.* AWWA B202-83, Denver, Colo.

Beals, J.L. (1976) Mechanics of Handling Lime Slurries. *Proc. Int. Water Conf.,* Pittsburgh, Pa.

Bitton, G., *et al.* (1980) *Sludge—Health Risks of Land Application.* Ann Arbor Science Publishers, Ann Arbor, Mich.

Burnham, J.C., *et al.* (1992) Use of Kiln Dust with Quicklime for Effective Municipal Sludge Treatment with Pasteurization and Stabilization with the N-Viro Soil Process. ASTM Stand. Tech. Publication 1135, Philadelphia, Pa.

Byers, H.W., and Jensen, B. (1990) Stabilizing Sludge with Fly Ash-Fludge. Paper presented at Dep. Eng. Prof. Develop., Univ. Wisconsin–Madison.

Christensen, G.L. (1982) Dealing with the Never-Ending Sludge Output. *Water Eng. Manage.,* **129,** 25.

Christensen, G.L. (1987) Lime Stabilization of Wastewater Sludges. In *Lime for Environmental Uses,* K.A. Gutschick (Ed.), ASTM, Philadelphia, Pa.

Christy, R.W. (1992) Process and Mechanical Design Considerations for Sludge/Lime Mixing. Paper presented at Water Environ. Fed. Spec. Conf. The Future Direction of Municipal Sludge (Biosolids) Management: Where We Are and Where We're Going, Portland, Oreg.

Counts, C.A., and Shuckrow, A.J. (1975) *Lime Stabilized Sludge: Its Stability and Effect on Agricultural Land.* EPA-670/2-75-012, Battelle Memorial Inst., Richland, Wash.

Engineering–Science, Inc., and Black and Veatch (1991) *Technology Evaluation Report: Alkaline Stabilization of Sewage Sludge.* Rep. prepared for U.S. EPA, EPA Contract No. 68-C8-0022, Work Assignment No. 01-08.

Farrell, J.B., *et al.* (1974) Lime stabilization of primary sludges. *J. Water Pollut. Control Fed.,* **46,** 113.

Fergen, R.E. (1991) Stabilization and Disinfection of Dewatered Municipal Wastewater Sludge with Alkaline Addition. Paper presented at Am. Water Works Assoc./Water Pollut. Control Fed. Joint Residuals Manage. Conf., Durham, N.C.

Jacobs, A., and Silver, M. (1990) Sludge Management at the Middlesex County Utilities Authority. *Water Sci. Technol.,* **22,** 93.

Jacobs, A., *et al.* (1992) Odor Emissions and Control at the World's Largest Chemical Fixation Facility. Paper presented at Water Environ. Fed. Spec. Conf. The Future Direction of Municipal Sludge (Biosolids) Management: Where We Are and Where We're Going, Portland, Oreg.

Kampelmacher, E.H., and van Noorle Jansen, L.M. (1972) Reduction of bacteria in sludge treatment. *J. Water Pollut. Control Fed.,* **44,** 309.

Lewis, C.J., and Gutschick, K.A. (1980) *Lime in Municipal Sludge Processing.* Natl. Lime Assoc., Washington, D.C.

Logan, T.J. (1990) Chemistry and Bioavailability of Metals and Nutrients in Cement Kiln Dust-Stabilized Sewage Sludge. Paper presented at Water Pollut. Control Fed. Spec. Conf., New Orleans, La.

Logan, T.L. (1990) Markets for N-Viro Soil. *Florida Water Res. J.,* 8.

Municipal Sewage Sludge Management: Processing, Utilization and Disposal (1992). C. Lue-Hing *et al.* (Eds.), Technomic Publishing Co., Inc., Lancaster, Pa.

National Lime Association (1988) *Lime: Handling, Application, and Storage in Treatment Processes.* Bulletin 213, Arlington, Va.

Oerke, D.W., and Rogowski, S.M. (1990) Economic Comparison of Chemical and Biological Sludge Stabilization Processes. Paper presented at Water Pollut. Control Fed. Spec. Conf. The Status of Municipal Sludge Management for the 1990s, New Orleans, La.

Otoski, R.M. (1981) *Lime Stabilization and Ultimate Disposal of Municipal Wastewater Sludges.* EPA-600/S2-81-076, U.S. EPA, Cincinnati, Ohio.

Paulsrud, B., and Eikum, A.S. (1975) Lime Stabilization of Sewage Sludges. *Water Res. (G.B.),* **9,** 297.

Ramirez, A., and Malina, J. (1980) Chemicals Disinfect Sludge. *Water Sew. Works,* **127,** 4, 52.

Reimers, R.S., *et al.* (1981) *Parasites in Southern Sludges and Disinfection by Standard Sludge Treatment.* EPA-600/2-81-166, U.S. EPA.

Rubin, A.R. (1991) Agricultural Limitations and Criteria for Lime Stabilized Sludges (PSRP or PFRP). Paper presented at the Am. Water Works Assoc./Water Pollut. Control Fed. Joint Residuals Manage. Conf., Durham, N.C.

Sloan, D. (1992) Design and Process Considerations for Advanced Alkaline Stabilization with Subsequent Accelerated Drying Facilities. Paper presented at 5th Annu. Int. Conf. Alkaline Pasteurization Stabilization, Somerset, N.J.

Stone, L.A., *et al.* (1991) Detailed Case Study Evaluation of Alkaline Stabilization Processes. Paper presented at the Am. Water Works Assoc./Water Pollut. Control Fed. Joint Residuals Manage. Conf., Durham, N.C.

Stone, L.A., *et al.* (1992) The Historical Development of Alkaline Stabilization. Paper presented at Water Environ. Fed. Spec. Conf. The Future Direction of Municipal Sludge (Biosolids) Management: Where We Are and Where We're Going, Portland, Oreg.

U.S. Environmental Protection Agency (1975) *Lime Stabilized Sludge: Its Stability and Effect on Agricultural Land.* EPA-670/2-75-012, Natl. Environ. Res. Center, Washington, D.C.

U.S. Environmental Protection Agency (1979) *Process Design Manual for Sludge Treatment and Disposal.* EPA-625/1-79-011, Cincinnati, Ohio.

U.S. Environmental Protection Agency (1989) *1988 Needs Survey of Municipal Wastewater Treatment Facilities.* EPA-430/09-89-001, Cincinnati, Ohio.

U.S. Environmental Protection Agency (1993) *Standards for the Use or Disposal of Sewage Sludge, Fed. Regist.,* **58,** 32, 40 CFR Part 503.

Water Environment Federation (1992) *Design of Municipal Wastewater Treatment Plants.* Manual of Practice No. 8, Alexandria, Va.; Am. Soc. Civ. Eng., Manual and Report on Engineering Practice No. 76, New York, N.Y.

Water Pollution Control Federation (1988) *Sludge Conditioning.* Manual of Practice No. FD-14, Alexandria, Va.

Westphal, A., and Christensen, G.L. (1983) Lime stabilization: effectiveness of two process modifications. *J. Water Pollut. Control Fed.,* **55,** 1381.

Chapter 7
Thermal Drying

Thermal drying involves the application of heat to evaporate water from sludge. Thermal drying reduces the moisture content to a level below that achievable by conventional mechanical dewatering methods. Compared with these methods, the advantages of thermal drying include reduced transportation costs, further pathogen reduction, improved storage capability, and marketability. Thermally dried biosolids can be easily marketed as a fertilizer or soil conditioner. They are acceptable for landfill disposal or efficient incineration.

The drying rate depends on the internal mechanism of liquid flow and the external mechanism of evaporation. During the drying process, a temperature gradient develops from the heated surface inward, causing moisture to migrate from within the wet sludge to its surface by the mechanisms of diffusion, capillary flow, and internal pressures generated by shrinkage during drying. These internal drying mechanisms must be understood to predict the probable behavior during the drying process. The transfer of heat from the heat transfer medium to the wet sludge raises the temperature and evaporates water at the solid surface. External conditions affecting the transfer of heat or the drying process include temperature, humidity, rate and direction of gas flow, the exposed surface area, physical form of sludge, agitation, detention time, and the method of sludge support used during drying. Understanding these external conditions and their effects is necessary when investigating the drying characteristics, choosing the correct dryer, and determining the optimal operating conditions. Although the internal and external mechanisms occur simultaneously, either mechanism may limit the drying rate.

*T*HERMAL DRYING STAGES

The three general stages of thermal drying, illustrated by Figure 7.1, include the warm-up stage, the constant-rate stage, and the falling-rate stage. Each is described below.

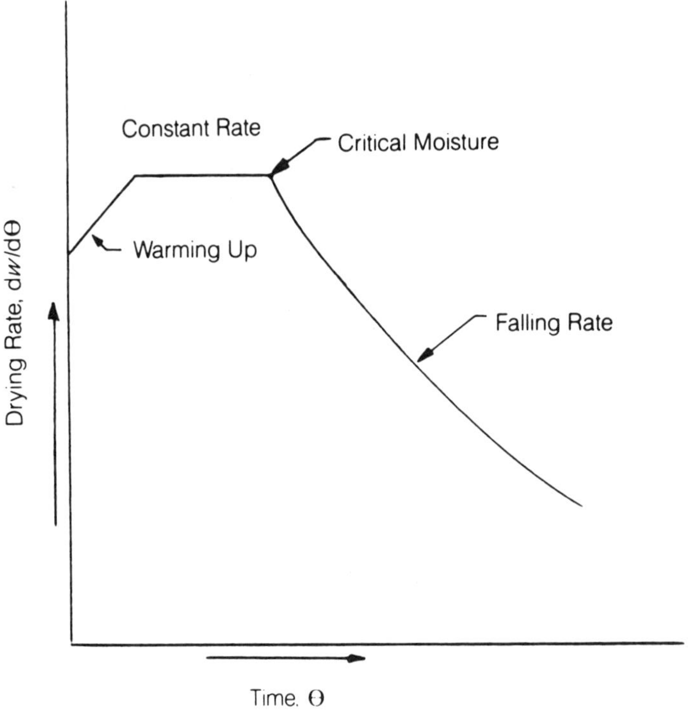

Figure 7.1   Three general stages of thermal drying (WEF, 1992).

WARM-UP STAGE. During the warm-up stage, the temperature and the drying rates increase to the steady-state conditions of the constant-rate stage. The warm-up stage typically is short and results in little drying.

CONSTANT-RATE STAGE. During the constant-rate stage, interior moisture replaces the external moisture as it evaporates from the saturated surface of the sludge. The transfer of heat to the evaporating surface controls the drying rate, similar to the evaporation of water from a pool of liquid. The constant-rate stage, typically the longest stage, results in most of the drying. The drying rate, essentially independent of the internal mechanism of liquid flow, depends on three external factors:

- Heat- or mass-transfer coefficient,
- Area exposed to the drying medium, and
- Temperature and humidity differences between the drying medium and the wet surface of the sludge.

FALLING-RATE STAGE. During the falling-rate stage, the external moisture evaporates faster than it can be replaced by internal moisture. As a result, the exposed surface is no longer saturated, latent heat is not transferred as rapidly as sensible heat is received from the heating medium, the temperature increases, and the drying rate decreases. The drying rate at the transition point between the constant-rate and falling-rate stages is called the critical moisture.

*H**EAT TRANSFER METHODS*

The classification of dryers is based on the predominant method of transferring heat. These methods are convection, conduction, radiation, or a combination of these.

CONVECTION. In convection (direct) drying systems, the wet feed directly contacts the heat-transfer medium, usually hot gases. Convection heat transfer is expressed mathematically as

$$q_{\text{conv}} = h_c A (t_g - t_s) \tag{7.1}$$

Where

q_{conv}	=	convective heat transfer, kJ/h (Btu/hr);
h_c	=	convective heat-transfer coefficient, kJ/m^2·h·°C (Btu/hr/sq ft/°F);
A	=	area of wetted surface exposed to gas, m^2 (sq ft);
t_g	=	gas temperature, °C (°F); and
t_s	=	temperature at sludge–gas interface, °C (°F).

The convective heat-transfer coefficient may be obtained from the dryer manufacturer or from pilot studies.

CONDUCTION. In conduction (indirect) drying systems, a solid retaining wall separates the wet sludge from the heat-transfer medium, typically steam or another hot fluid. The mathematical expression for conduction heat transfer is

$$q_{\text{cond}} = h_{\text{cond}} A(t_m - t_s) \tag{7.2}$$

Where

q_{cond}	=	conductive heat transfer, kJ/h (Btu/hr);
h_{cond}	=	conductive heat-transfer coefficient, kJ/m²·h·°C (Btu/hr/sq ft/°F);
A	=	area of heat-transfer surface, m² (sq ft);
t_m	=	temperature of heating medium, °C (°F); and
t_s	=	temperature of sludge at drying surface, °C (°F).

This equation assumes 100% of the area of heat-transfer surface in contact with the feed. The conductive heat-transfer coefficient (h_{cond}), a composite term, includes the effects of the heat-transfer surface films of the feed and the heating medium. The coefficient may be obtained from the dryer manufacturer or from pilot studies.

RADIATION. In radiation (infrared) drying systems, infrared lamps, electric resistance elements, or gas-heated incandescent refractories supply radiant energy that transfers to the wet feed and evaporates moisture. Radiation heat transfer is expressed as

$$q_{\text{rad}} = C_s As (t_r^4 - t_s^4) \tag{7.3}$$

Where

q_{rad}	=	radiation heat-transfer, kJ/h (Btu/hr);
C_s	=	emissivity of the drying surface, dimensionless;
A	=	sludge surface area exposed to radiant source, m² (sq ft);
s	=	Stefan–Boltzman constant, 4.88×10^{-8} kcal/m²·h·°K (1.73×10^{-9} Btu/hr/sq ft/°R);
t_r^4	=	absolute temperature of radiant source, °K (°R); and
t_s^4	=	absolute temperature of radiant source, °K (°R).

The emissivity of the drying surface may be obtained from the dryer manufacturer or from pilot studies.

Wastewater Residuals Stabilization

PROCESS DESCRIPTION

Thermal dryers are grouped into four categories: direct, indirect, combined direct–indirect, and infrared. Each is discussed below.

DIRECT DRYERS. Direct (convection) dryers that have been successfully used include the flash dryer, rotary dryer, and fluid bed dryer.

Flash Dryer. The flash dryer or pneumatic conveyor dryer (Figure 7.2) consists of a furnace, mixer, cage mill, cyclone separator, and vapor fan (U.S. EPA, 1979). The mixer blends wet feed with dried product to achieve a mixture with 40 to 50% moisture (Metcalf and Eddy, Inc., 1979, and Niessen, 1988). The feed mixture discharges to a windbox on the cage mill. In the windbox, hot furnace gases at a temperature of 650 to 705°C (1 200 to 1 300°F) disperse the feed and rapidly evaporate moisture (U.S. EPA, 1979). The cage mill mechanically agitates the sludge–gas mixture to maximize the surface area contacting the hot gases and complete the drying process. The dry biosolids, with a moisture content of 8 to 10%, is pneumatically conveyed to a cyclone separator that separates the biosolids from the spent drying gases (U.S. EPA, 1979). The dried biosolids may be marketed as a fertilizer or soil conditioner. The product may also be mixed with wet feed and incinerated autogenously in the furnace to reduce auxiliary fuel requirements.

Flash-dried biosolids are extremely dusty, creating the potential for fires and explosions. Dust may also complicate handling, storage, and marketing. Pelletization of the dried biosolids can improve marketability, but at additional capital and operations and maintenance expense. Grease buildup in pipelines and ducts can also cause fires and explosions. The flash dryer, a complex system with several heat exchangers and numerous material handling processes, is vulnerable to severe abrasion by dried biosolids (especially the cage mill and cyclone separator) and fouling of the heat exchanger surfaces. Another disadvantage of flash dryers is their relatively high operating cost. As with other dryers, the system economics will improve if some of the biosolids can be marketed as a fertilizer or soil conditioner.

Design considerations for a flash dryer system include

- Moisture content—maintain the moisture content of the feed below approximately 40 to 50% to ensure that the gas stream will readily disperse the feed without agglomeration and sticking on the conveyor walls (Williams–Gardner, 1982).
- Particle size—detention time in the flash dryer typically averages only a few seconds. Hence, the drying process is mainly a surface phenomenon that demands small particles to allow essentially instantaneous heat and mass transfers from interiors to exteriors.

- Gas velocity—the gas velocity sufficient to carry the largest particle to the cyclone separator must typically range from 20 to 30 m/s (65 to 100 ft/sec) (U.S. EPA, 1979).

- Agitation—the cage mill must agitate the feed sufficiently to break up small clumps, thereby exposing new surface areas and creating enough turbulence for dispersal and entrainment of particles in the hot gas stream.

- Air pollution control—cool gases from the cyclone are preheated to approximately 595°C (1 100°F) through heat exchange with combustion gases from the furnace. Then the cyclone gases, mixed with

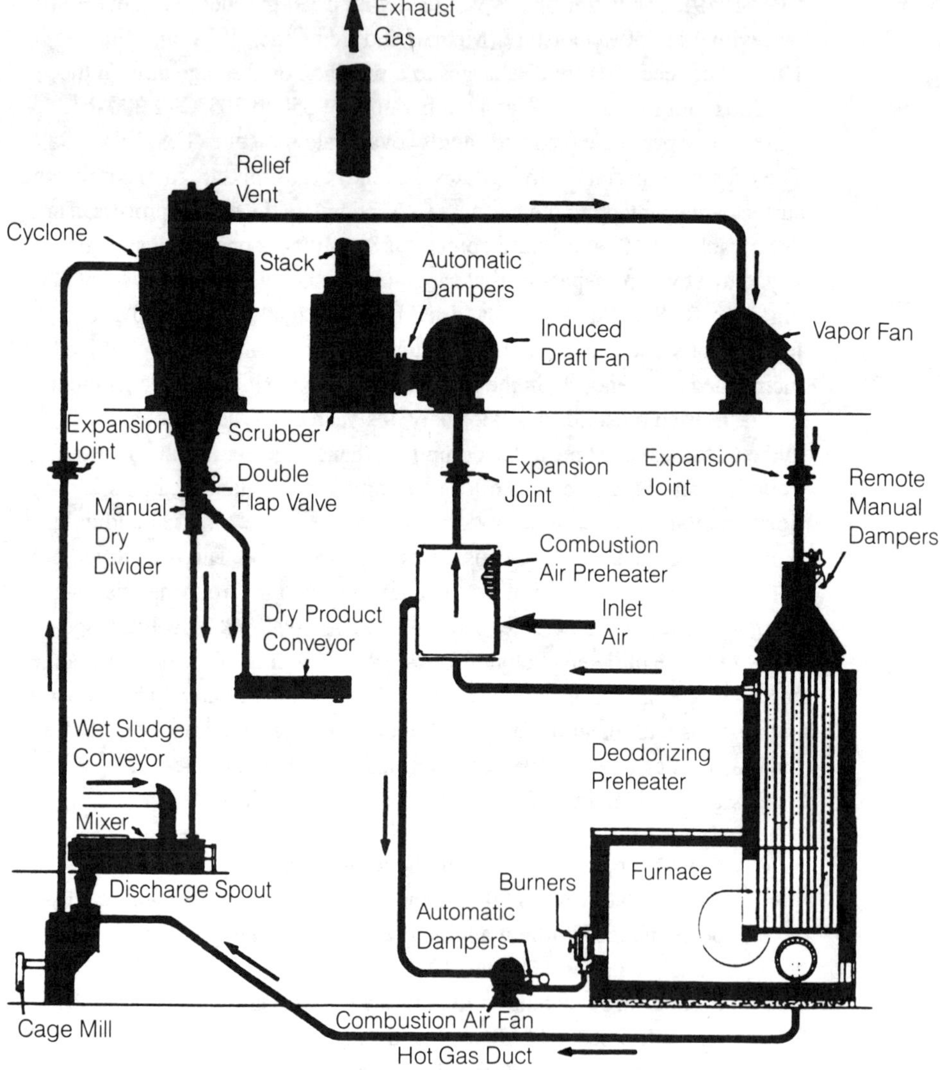

Figure 7.2   Flow diagram for flash dryer system (WEF, 1992).

the combustion gases to attain a minimum temperature of 650°C
(1 200°F), pass through the deodorizer preheater to the wet scrubber
for particulate removal (Niessen, 1988). Preheater temperatures of
760°C (1 400°F) or higher may be required for effective odor control. Exhaust gases from the wet scrubber may require further treatment to reduce air emissions before exiting the stack.

Rotary Dryer. The rotary dryer consists of a cylindrical steel shell, rotated
on bearings, typically mounted with its axis at a slight slope from the horizontal
(Figure 7.3). Rotary dryers include both single-pass and triple-pass systems.

A schematic diagram of the rotary dryer system is shown in Figure 7.4. A
mixer blends wet feed with previously dried biosolids, thereby reducing the
moisture content of the feed material to 30 to 40% and dispersing the cake
and beginning the formation of pellets (Niessen, 1988). The blended feed continuously enters the upper end of the rotary dryer along with hot furnace
gases at a temperature ranging from 260 to 480°C (500 to 900°F) (Niessen,
1988, and U.S. EPA, 1979). The feed mixture and hot gases are conveyed,
typically cocurrently, to the discharge end of the dryer. During conveyance,
axial flights along the slowly rotating interior wall of the dryer pick up and cas-
cade the feed through the dryer. This creates a thin sheet of falling particles,

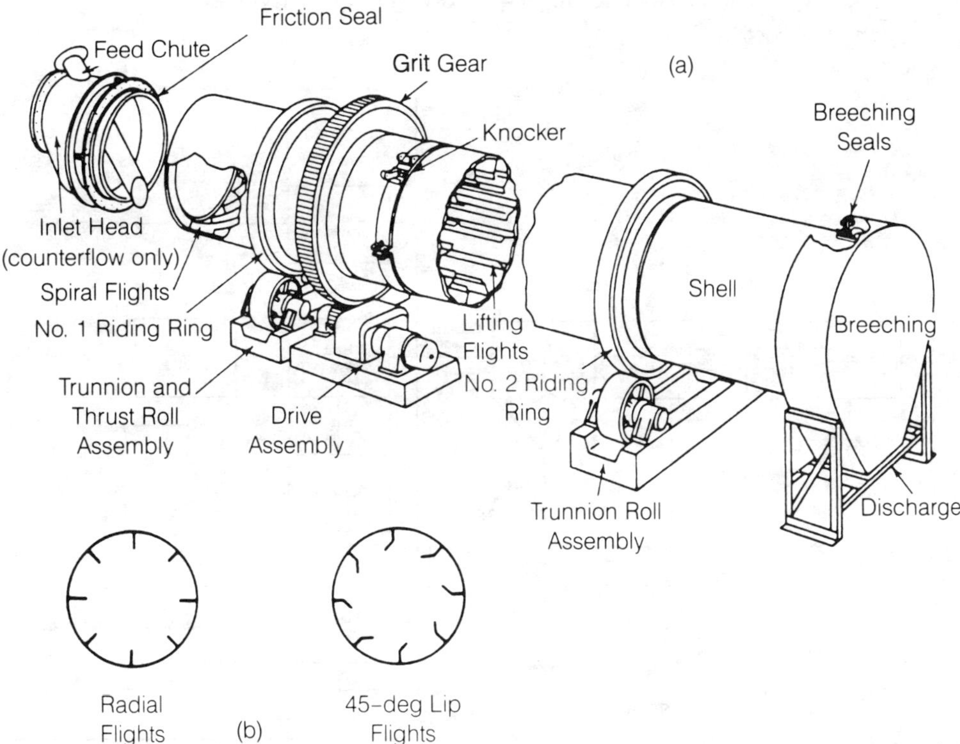

**Figure 7.3   Rotary dryer: (a) isometric view and (b) alternative flight arrangements
(WEF, 1992).**

---

which directly contact the hot gases and dry rapidly. The exhaust gases exit
the dryer at 65 to 105°C (150 to 220°F) and travel to air pollution control
equipment for odor and particulate removal (Niessen, 1988).

Rotary dryers have successfully dried raw primary and secondary mixtures, waste-activated sludge, and digested primary biosolids. The dried
biosolids product is amenable to handling, storage, and marketing as a fertilizer
or soil conditioner.

Design considerations for a rotary dryer include

- Moisture content—maintain the moisture content of the feed below
  30 to 40% by blending previously dried biosolids with the wet feed.
  This will ensure that particles do not stick to the conveyor, flights,
  and interior dryer walls.
- Length of drum—the length of the rotary drum typically ranges between 4 and 10 times the drum diameter for single-pass dryers and 2.5
  and 3 times the drum diameter for triple-pass dryers (Perry and Green,
  1984).
- Drum rotation and slope—the rotary drum typically rotates at 5 to
  8 rpm for single-pass dryers and 10 to 12 rpm for triple-pass dryers
  (U.S. EPA, 1979). The drum slope typically is 2 to 6 cm/m (0.25 to
  0.75 in./ft) of drum length for single-pass dryers. Triple-pass dryers
  are not sloped.

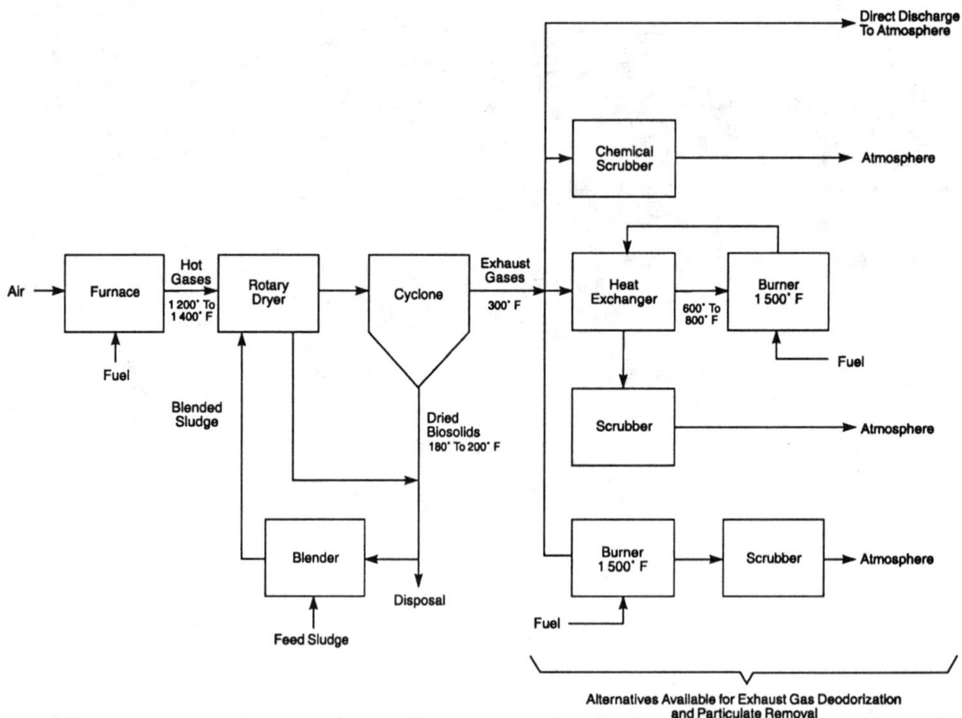

Figure 7.4   Flow diagram for a rotary dryer system (WEF, 1992).

- Flight spacing—axial flights in the interior of the dryer typically are offset every 0.6 to 2 m (2.0 to 6.6 ft) (Perry and Green, 1984).
- Flight geometry—the design of the axial flights depends on the handling characteristics of the material. Many dryers are designed with flat radial flights without any lips for the first 30% of the dryer and with flights with 45- to 90-deg lips for the last 65% of the dryer. Sometimes with cocurrent gas flow, flights are omitted in the last meter of the dryer to reduce the entrainment of dried particles in the exit gas.
- Gas velocity—gas flow through the drum may be either cocurrent or countercurrent to the feed flow. Gas velocities typically are limited to 1.2 to 3.7 m/s (4 to 12 ft/sec); otherwise, excessive amounts of dust will exit the dryer (U.S. EPA, 1979). Gas velocity must also be controlled to obtain an adequate gas retention time and efficient heat transfer between the gas and feed.
- Air pollution control—air pollution control typically includes treatment for particulate removal and for odor control by chlorine or permanganate-based scrubbers or by thermal oxidation (afterburners).

Fluid Bed Dryer. The fluid bed dryer consists of a stationary, vertical chamber with a perforated bottom through which hot gases, typically air, are forced or induced by a blower arrangement. Figure 7.5 shows a typical diagram of a fluid bed dryer system.

A screw conveyor carries the wet feed through an air lock. In the dryer, the heated gas fluidizes or suspends the feed and sand bed. The heated gases from the system furnace enter the plenum chamber and a perforated plate at the base of the dryer evenly distributes the gases across the body of the dryer. This produces a high level of intermixing and an intimate gas–solids contact, resulting in high mass and heat transfer between the solid and gas phases. Dried biosolids exit the dryer by overflowing from the chamber through a pipe or adjustable weir with a rotary air lock. Vent gases exhaust to a cyclone separator or other air pollution control equipment.

Although the fluid bed dryer has been proven to be an efficient drying method for many applications, it is seldom used for drying municipal wastewater residuals. A fluid bed dryer installation in Norwalk, Connecticut, was used prior to fluid bed combustion. The dryer operated without problems, approximately doubling the capacity of the combustion unit, before it was shut down because of problems associated with the 25-year-old fluid bed furnace.

INDIRECT DRYERS. Indirect (conduction) dryers that have been used for drying municipal residuals include the paddle dryer, hollow-flight dryer, disc dryer, and multiple-effect evaporation dryer.

Paddle, Hollow-Flight, or Disc Dryer. The paddle, hollow-flight, or disc dryer consists of a stationary horizontal vessel or trough with a jacketed shell through which a heat-transfer medium (typically steam) circulates. The vessel or trough contains a rotating agitator assembly with a series of agitators (discs, flights, or paddles) mounted on a rotating shaft (rotor). The rotor and agitators (typically hollow) allow the heat-transfer medium to circulate through the hollow core. Therefore, the agitators not only transport the feed through the unit, but also provide an additional heat-transfer surface that contacts the feed. A schematic diagram of a paddle or hollow-flight dryer is shown in Figure 7.6.

Wet feed flows continuously into the vessel or trough, and the agitator surface, pitched at a slight angle, conveys the wet feed through the dryer. In some dryers, the agitators are arranged on the shaft to intermesh, thereby producing the maximum possible relative movement between the heated surfaces and the feed to enhance drying. The meshed agitators increase the heat-transfer coefficient and self-clean the agitator surface. They are susceptible to damage, however, from physical contaminants (for example, metal fragments) that may be contained in the feed. In other dryers, stationary agitator ploughs or breaker bars are located between the rotating agitators to achieve better mixing and prevent solids buildup on the agitator surface. The heat-transfer medium circulates, typically countercurrent, through the jacketed shell and hollow agitator assembly. The transfer of heat raises the temperature and evaporates water from the solid surface. The evaporated water is transported

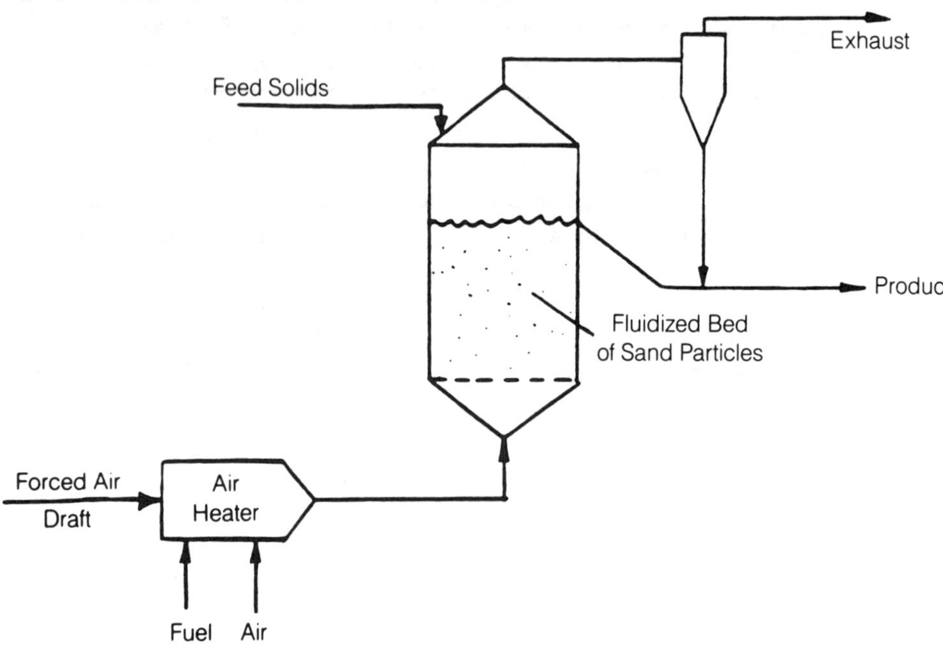

Figure 7.5   Flow diagram for a fluid bed dryer system (WEF, 1992).

from the surface of the solids and out of the dryer by low-volume sweep gases or exhaust vapors. A weir at the discharge end of the dryer ensures complete submergence of the heat-transfer surface in the material being dried.

Design considerations for paddle, hollow-flight, or disc dryers include

- Moisture content—the moisture content of the feed typically decreases to 40 to 50% because previously dried biosolids are blended with wet feed to prevent agglomeration and fouling of the dryer surfaces (Niessen, 1988). If the feed solids content is in the tacky plastic stage, clumps may form and agitators may become plated. Some manufacturers claim to be capable of drying even in the plastic stage without backmixing dried product. This approach must provide enough power to turn the agitator shaft to break up the clumps against internal breaker bars and must also provide a strong enough agitator shaft to avoid damage.
- Agitation—agitation improves the rate of heat transfer, thus the rate of drying. As the degree of agitation increases, dependence on the thermal conductivity of the process mass is reduced. As more individual

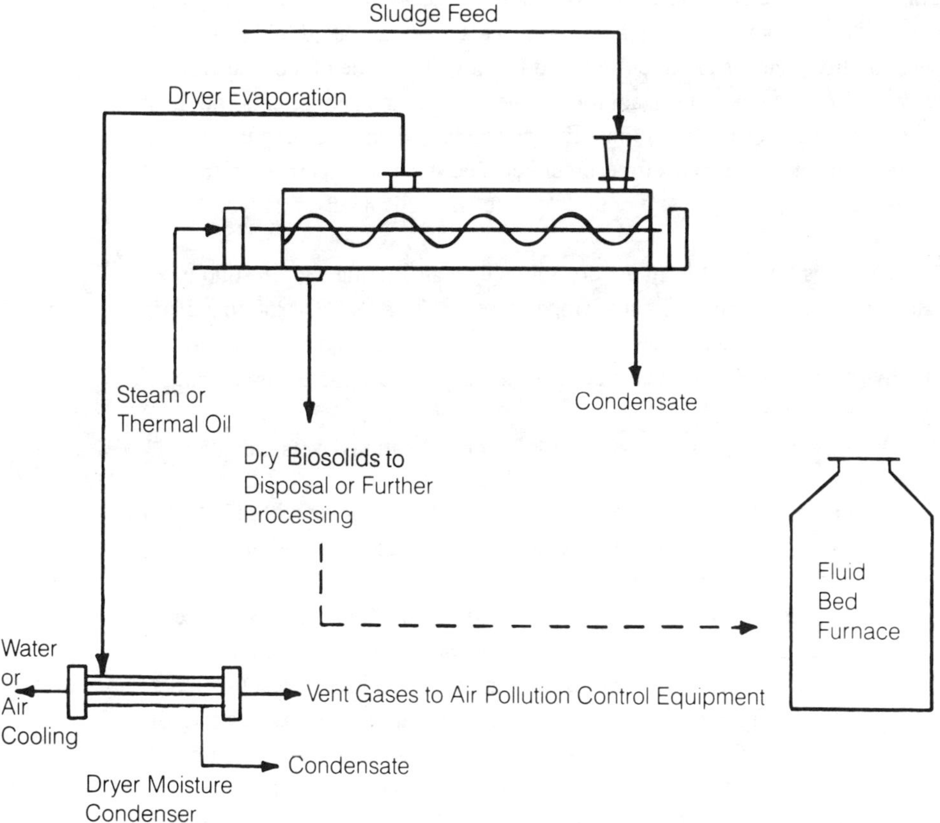

Figure 7.6　Flow diagram for a paddle or hollow-flight dryer system (WEF, 1992).

---

particles of mass are exposed to the heat-transfer surface, bed temperatures within the dryer become more uniform.

- Gas velocity—the paddle, hollow-flight, or disc dryer typically uses little or no sweep gas. When used, the sweep gas velocity typically is controlled at approximately 0.15 m/s (0.5 ft/sec).

Emerging Technology. The multiple-effect evaporation process extracts water by evaporation. The four major steps in the process are oil mixing, multiple-effect evaporation, oil–solid separation, and condensate–oil separation. A grinder macerates dewatered cake (approximately 20% solids) to reduce the size of the cake particles, thereby preventing clogging of the evaporator tubes, optimizing evaporation, and simplifying control. The macerated cake is mixed with a carrier oil, such as No. 2 fuel oil in the fluidizing tank at a suggested ratio of 1 part dry solids to 5 to 10 parts oil by weight. The carrier oil reduces scaling, inhibits corrosion, and enhances heat transfer, while keeping the cake–oil slurry fluidized.

The slurry is pumped to the first-effect evaporator and steam is added to the shell side of the evaporator at a minimum of 345 kN/m^2 (50 psig) (U.S. EPA, 1979). The vaporized water is then removed as the cake–oil slurry rolls down the evaporator tubes in a thin film. This action is called falling-film evaporation. The water vapor removed from the tube side of the evaporator (first effect) provides the steam for the shell side of the next effect. The steam used on the shell side of the first effect is condensed and returned to the boiler. The process repeats for each effect. The steam downstream of the first evaporator is condensed and returned to the hot well and eventually to the sewer.

The oil is separated from the dry solids by centrifuging the oil–solid mixture and by hydroextraction at a temperature of 120 to 160°C (250 to 320°F) at atmospheric pressure. Dry biosolids product discharges from the system after hydroextraction and process condensate recycles back to the treatment plant.

Design considerations for multiple-effect evaporation dryers include

- Oil recovery—cost-effective operation of the process with light fluidization oil requires recovery and reuse of the light oil to reduce the cost of light oil makeup.
- Number of effects—through the reuse of heat in the multiple-effect process, the amount of water removed per kilogram of steam supplied increases as the number of effects increases. In its simplest theoretical form, a single-effect evaporator can evaporate a maximum of 1 kg of water/kg of steam supplied (lb/lb). A double-effect evaporator will evaporate 2 kg of water/kg of steam supplied through the reuse of heat. Theoretically, an infinite number of effects would result in an infinite economy of steam. Several factors, of course, limit the practical

number of effects in a system. Each effect of a multiple-effect evaporator operates with only a fraction of the total temperature drop across the system. The total drop for a multiple-effect system is seldom larger than the drop for single-effect evaporation. The capacity per unit area of heating surface for a multiple-effect system is reduced proportionately. Thus, the savings in fuel requirements that are achieved through multiple-effect operation are offset by increased equipment costs. In most cases, no more than three or four effects are economical. The actual number of economical effects is largely influenced by prevailing fuel costs.

- Air pollution control—vent gases and odorous emissions from the dehydration process can be collected and eliminated by combustion in the boiler or furnace.
- Staffing requirements—process requires some of the same equipment and operator skills found in the petrochemical industry. Hence, the drying facility will typically have to pay higher wages and provide more training than would be required for conventional wastewater treatment plant operation.

DIRECT–INDIRECT DRYERS. The direct–indirect dryer, a jacketed trough or vessel, uses larger volumes of hot gases (typically air) as a heating medium or sweep gas than does the indirect dryer. This mode of operation lowers the boiling point of the volatile fraction and causes a fluidizing effect, therefore improving the heat-transfer coefficient between the feed and heating surfaces. Another advantage of this mode of operation is that the direct and indirect heating can be varied to reduce energy consumption and increase the drying rate.

INFRARED DRYERS. Municipal applications of infrared (radiant) dryers typically involve combustion (that is, an infrared furnace or multiple-hearth furnace). The most important design considerations for thermal drying systems are briefly discussed below. For more information, consult *Design of Municipal Wastewater Treatment Plants* (WEF, 1992).

DRYER SELECTION. Indirect dryers typically are the best selection for dryers preceding combustion, especially where steam or exhaust heat may be recovered from the combustion process. Because indirect dryers generate limited quantities of noncondensable gas, little odor control treatment is required. Likewise, dust problems are reduced because of the small volume of carrier or sweep gases used in indirect drying. Indirect dryers allow operation under a vacuum or closely controlled atmosphere. Therefore, fire and explosion hazards are reduced within such units.

A disadvantage of indirect dryers is that, while dusting is reduced during the drying process, the dried biosolids product tends to be dustier than the direct rotary dried product. This may limit marketing options, increase the cost of dried product handling and storage, and necessitate further processing of the product by compaction or other granulating technologies to create a product acceptable to the fertilizer industry. Hence, for drying applications where the production of a fertilizer or soil amendment is desired, direct dryers typically are the best choice. Direct dryers have historically produced a marketable biosolids product.

If the goal is to achieve a final product solids content of 65 to 85%, then indirect dryers offer a higher thermal efficiency (ratio of heat input to heat required for liquid evaporation) than do direct dryers. However, to achieve a solids content of 92% with indirect dryers for safe product storage, a higher temperature must be used, drying time must be increased, or dried biosolids must be recycled and mixed with the feed. These steps will reduce the thermal efficiency advantage of indirect dryers compared with direct dryers.

MOISTURE CONTENT. Moisture content, an important characteristic, affects material handling and dryer capacity. Mechanical dewatering methods (that is, centrifuges and filter presses) typically are more efficient per volume of water removed than are thermal drying methods. Mechanical methods, however, are unable to obtain the low moisture levels attainable with thermal drying methods. As a result, mechanical dewatering before thermal drying deserves consideration.

Often, thermal drying requires backmixing dried product to achieve a solids content above the plastic stage to prevent agglomeration and the fouling of internal dryer surfaces associated with drying in the plastic stage. Because backmixing requires a larger dryer capacity, the amount of dried product that is backmixed must be considered in the system design.

BACKMIXING EQUIPMENT. Paddle mixers, pug mills, and hammer mills are sized to thoroughly mix recycle and feed. For backmixing, heavy-duty equipment is needed to reduce imbalance and bearing problems that result from fouling.

FEEDER SELECTION. Wet cake requires uniform feeding to maintain effective drying. A self-cleaning, intermeshed twin-screw feeder will effectively control solids buildup on the screw conveyors.

CONSTRUCTION MATERIALS. The composition of the feed significantly affects the construction materials selection for a dryer and ancillary equipment. Although mild steel typically will be appropriate for components that contact dry solids, stainless steel or other corrosion-resistant materials may be required for components in contact with wet feed. The abrasiveness of the feed calls for consideration of hard surfacing of areas especially prone to wear (for example, agitators in indirect dryers) to avoid frequent equipment replacement. Dryer and ancillary equipment need to be airtight and well insulated to reduce heat losses and consequent loss of efficiency.

STORAGE. Storage, an important and often overlooked system element, typically precedes and follows thermal drying. Sufficient storage is needed ahead of the drying system to allow post-drying-system shutdowns and attenuate variations in plant solids production. Also necessary, postdrying storage should be sufficient to handle seasonal fluctuations in schedules for hauling or incineration. Typically, between 30 and 90 days of product storage is a reasonable allowance for drying and marketing operations.

Typically, the dried product is stored in silos, sealed from the outside environment. Operating experience has shown that stored product with a moisture content greater than 10% can spontaneously combust. This fire hazard can be reduced by the addition of nitrogen gas to the storage silo to create an oxygen-free environment; this can be costly, however. Pelletization before storage will reduce the fire hazard by reducing the dust associated with the dried product. The best method of preventing spontaneous combustion entails drying to achieve 92% or more solids. As good practice, the design engineer provides for explosion safety hatches to release pressure in the storage bins as necessary.

EXPLOSIVE HAZARD REDUCTION. Dried biosolids produce a fine organic dust, which can reach explosive concentrations. The hazard is reduced by keeping the dryer temperature less than approximately 540°C (1 000°F). The dryer and dried biosolids conveying system from the process air streams may be designed to control dust by maintaining a negative pressure in dust-prone areas. Also, high-efficiency cyclones, Venturi scrubbers, bag houses, or a combination of these can be provided to remove dust and fine particles from the process air streams. Well-trained operators, controls with interlocks, and monitoring equipment are important considerations for reducing the fire hazard. Automated water spray systems can control fires with little or no damage to equipment. A centralized fire extinguisher system that will inject carbon dioxide into the dryer or a water dousing system are other design considerations for controlling fires. As for fire control in storage bins, a nitrogen gas blanket and product precooling can reduce the fire hazard.

AVAILABLE SPACE. Site conditions, including space and height available for the drying system, are considerations for the selection and design of the dryer and ancillary equipment.

EMISSIONS AND ODOR CONTROL. As good design practice, dryer equipment and handling and storage areas are contained and vented to air pollution control equipment. Chemical scrubbing and afterburners typically control odors. Cyclone separators, wet scrubbers, bag houses, or a combination of these remove particulates.

SIDESTREAMS. Odorous liquid sidestreams produced by condensation of water vapors from dryers contain both organic oils and ammonia. Wet scrubbers and other ancillary equipment will generate other sidestreams. The effects of all sidestreams on the wastewater treatment plant deserve the design engineer's consideration.

SAFETY. The design considerations for a thermal drying system include an emphasis on system safety. In addition to the safety considerations listed for reducing fire hazards, continuous monitoring of important operating variables such as temperature, feed rates, air rates, and drying time is required. Instrumentation is best positioned where operators may easily observe and respond to process indicators or alarms. An automatic shutoff, as a backup to the alarm system, is necessary in the system design.

HEAT RECOVERY. Dryer design provides for heat recovery and reuse wherever feasible. Recovered heat from dryer or furnace exhaust may be reused to preheat combustion air, preheat feed, or supplement plant heating requirements.

LABOR REQUIREMENTS. The complexity of some municipal drying systems necessitates specialized staffing during all operating hours. Often, drying facilities have failed because the municipality or owning authority failed to provide highly qualified operators or process engineers. Some municipal treatment plants now contract with private firms to operate drying and marketing operations.

SYSTEM ECONOMICS. Fuel consumption is the major component of operating costs. Alternative fuels can reduce this cost considerably. Potential fuel sources for the system include natural gas, methane gas from landfills or digesters, steam, or waste heat from cogeneration or incineration, wood chips, and the dried biosolids product. If a sound marketing approach is used, the operational costs of the process can be offset or partly offset by the return on product sales.

PRODUCT MARKETING. All successful biosolids marketing operations have recognized that revenues earned from the sales of dried product will not cover the full cost of processing and marketing the products. These operations are nonetheless successful because revenues from the sales and distribution of the biosolids product reduce overall net solids processing costs. Successful marketing requires a distribution system not vulnerable to extreme fluctuations in demand. Municipal biosolids typically do not have a high nutrient content, but the high organic content (about 65 to 85% organic by dry

---

weight) serves well in soil conditioning and retaining soil moisture. Most biosolids programs blend the biosolids with fertilizer materials high in nutrient content to further increase the usefulness and value of the product. Important marketing and pricing needs for biosolids include

- Nitrogen content—the biosolids need at least 3 to 4% nitrogen to be acceptable for direct application as a fertilizer (Schafer and Mathis, 1987). Biosolids with lower nitrogen contents may be used as constituents of blended fertilizers, but have lower value.
- Moisture content—the moisture content must be 10% or less to meet the EPA definition for heat drying as a process to further reduce pathogens and to minimize the chance of spontaneous combustion (U.S. EPA, 1994).
- Pathogen free—to meet U.S. EPA criteria for a pathogen-free product, the biosolids particles must reach or exceed dry bulb temperature of 80°C (175°F) and the exhaust gas must reach or exceed a wet bulb temperature of 80°C.
- Particle size and uniformity—the size of the dried biosolids particles must suit the intended user. A poorly graded product might result in a reduced price, a limited market, and rejection by brokers and users.
- Durability—particles must be strong enough to withstand handling and transportation with limited breakage.
- Dustless product—the dried biosolids must be as dust-free as possible for ease of handling and application.
- Ability to dissolve in soil—the dried biosolids particles must dissolve in soil over time to release nutrients into solution for plant uptake.
- Regulations—U.S. EPA 40 CFR Part 503 regulations establish minimum federal requirements for final use and disposal of wastewater biosolids (U.S. EPA, 1993). These regulations limit the annual loadings of heavy metals and toxic organic contaminants for land application of biosolids and blended biosolids-based fertilizers. In addition, 40 CFR Part 503 establishes criteria for pathogen and vector attraction reduction.
- Odor free—the dried biosolids product must be as odorless as possible for handling, storage, and application.
- Free of extraneous materials—the dried biosolids should be free of extraneous materials (such as plastics and string).
- Reliability of supply—the supply of dried biosolids must be reliable to achieve customer loyalty. Name recognition is also important in this regard.
- Price—the price that dried biosolids may command is based on the nitrogen content. Unit prices in recent years have been in the range of $10 to $20 per each percentage point of nitrogen per ton of material.

PILOT TESTING. Many manufacturers have a laboratory for demonstrating their drying equipment with a customer's feed material. In addition, some system suppliers provide demonstration equipment for actual plant tests. Drying equipment successfully demonstrated under conditions approaching those of actual service might be more dependable than other equipment lacking such testing.

*R*EFERENCES

Metcalf and Eddy, Inc. (1979) *Wastewater Engineering Treatment, Disposal and Reuse.* McGraw–Hill, Inc., New York, N.Y.

Niessen, W.R. (1988) Thermal Processing of Wastewater Treatment Plant Sludges. *Proc. Natl. Conf. Munic. Sew. Treatment Plant Sludge Manage.,* Palm Beach, Fla.

Perry, R.H., and Green, D.W. (Eds.) (1984) *Perry's Chemical Engineers' Handbook.* McGraw–Hill, Inc., New York, N.Y.

Schafer, P., and Mathis, A. (1987) Making Dried Sludge Marketable. Paper presented at Water Pollut. Control Fed. Spec. Conf., Los Angeles, Calif.

U.S. Environmental Protection Agency (1979) *Process Design Manual for Sludge Treatment and Disposal.* EPA-625/1-79-011, Washington, D.C.

U.S. Environmental Protection Agency (1993) *Standards for the Use or Disposal of Sewage Sludge. Fed. Regist.,* **58,** 32, 40 CFR Part 503.

U.S. Environmental Protection Agency (1994) *A Plain English Guide to the EPA Part 503 Biosolids Rule.* Office of Wastewater Manage., Washington, D.C.

Water Environment Federation (1992) *Design of Municipal Wastewater Treatment Plants.* Manual of Practice No. 8, Alexandria, Va.; Am. Soc. Civ. Eng., Manual and Report on Engineering Practice No. 76, New York, N.Y.

Williams–Gardner, A. (1982) *Industrial Drying.* Gulf Publishing Company, Houston, Tex.

Chapter 8
Emerging Technologies

With increased public concern about environmental issues in the U.S., there is an interest in the disposition of municipal biosolids with political, technical, economical, and public health ramifications. This chapter describes and discusses the feasibility of some of the emerging treatment processes recently introduced for resource or byproduct recovery and disposition. The importance of this subject matter can be attributed to public opinion expressed about current biosolids disposal methods such as landfill dumping, artificial wetlands soils, and reuse on farm lands (Whittington and Johnson, 1985). It has been clearly mandated that new and innovative treatment processes must be introduced to provide a safe and economical product suitable for a diversity of reuse modes dependent on local needs.

To resolve the question of reusing municipal biosolids, it is first necessary to determine if the management and treatment procedures are in accordance with existing regulatory guidelines, are cost effective, and are compatible with the environment concerned.

On February 19, 1993, the Final Rule *Standards for the Use or Disposal of Sewage Sludge* regulations were published by the U.S. Environmental Protection Agency (U.S. EPA, 1993). As a result of new laws, biosolids process schemes are being developed with the use of treated municipal biosolids in mind. The possible end uses of the biosolids are listed below:

- Agricultural (amendment, liming agent, or fertilizers),
- Building fill,
- Landfill cover,
- Erosion control,
- Landfill leachate treatment,
- Road bed material,
- Fuel source,
- Building materials (such as bricks), and
- Artificial wetland soils (amendment or fertilizers).

The characteristics of the treated biosolids product must fit the requirements of the end use. In addition, the treated biosolids must not release toxics at levels affecting humans or the environment, contain infectious agents, or cause nuisances to the public.

CURRENT EMERGING TECHNOLOGIES

The definition of an emerging technology is a process that has been applied on a pilot or mobile scale and has a potential for application in wastewater treatment plants (WWTPs). In addition, these processes have exhibited both technical and economical advantages over conventional processes with potential applications within the next few years. The processes may have been used in major industrial settings, but not in the treatment of municipal biosolids.

As a result of the necessity for the beneficial use of municipal biosolids instead of disposal, the subject of successful disinfection and stabilization has arisen because biosolids from municipal WWTPs may contain pathogens hazardous to humans and domestic animals. The U.S. EPA Final Rule *Standards for the Use or Disposal of Sewage Sludge* regulations have established criteria for pathogen control and disposition limitations along with the appropriate stabilization treatment. Among the emerging biosolids disinfection processes that have developed in the last 5 years are

- Thermophilic pozzolanic fixation processes,
- Acid oxidation/disinfection processes,
- Heat treatment/acid digestion processes,
- Heat/wet air oxidation processes, and
- Active sludge pasteurization processes.

All these processes have operated at the demonstration field pilot stage
and are applicable in the municipal setting depending on specific site situations.
In addition, these processes may yield a Class A pathogen-free product under
the U.S. EPA Final Rule *Standards for the Use or Disposal of Sewage Sludge*
regulations because of equivalence testing and operating conditions.

THERMOPHILIC POZZOLANIC FIXATION PROCESSES. These
processes use alkaline agents such as limes and pozzolanic cements in conjunction with pozzolanic accelerators based on soluble sodium silicate. This
technology is a relatively simple and inexpensive process; has a mobile for-
mat that allows rapid setup; and the cured product is stable, noninfectious,
and nontoxic.

Process Description. Through the use of specially designed chemical reagents, rapid chemical heat is used in combination with a rapid rise in alkalinity and the evolution and control of ammonia to produce a fast-curing,
soillike product. This product is pathogen-free and has immobilized toxic
metals that might have been present in the raw sludge.

The process contains three key mixing components: two pug mills and one
screw feed conveyor. In smaller throughput arrangements, one pug mill may
suffice. All components may be constructed of mild steel.

Each mixer is totally enclosed so that all of the vapors generated as the
chemical reactions proceed are contained. These mixers and the screw conveyor are vented into a plug-flow retention vessel that allows the chemical reactions to proceed under optimal thermal conditions, as well as a vessel for
the controlled release and management of ammonia. If sustained downtime is
required for the overall process, this vessel would have to be manually
cleaned through the dual hatches.

Other than the exhaust gases (which are scrubbed), there are no sidestreams resulting from this process. The system uses the chemical processing
power inherent in thermophilic pozzolanic chemistries. The essential reactions
are shown in Figure 8.1.

The catalytic co-reactant for pozzolanic chemistries is calcium hydroxide.
This chemical, when reacted with aluminum, silicon and iron oxides, and poz-
zolan oxide complexes, releases significant heat. This is added to the heat
generated in the formation of calcium hydroxide from the oxide form.

In addition to the initial heat of reaction (IHR) effects on the temperature
of this product, this process takes advantage of the effect the IHR has on the
secondary and tertiary reactions that immediately follow in pozzolanic chemistry.

As the heat begins to build, there is an increased efficiency in the chemical
reactions, leading to the generation of additional heat from all of the chemical
reactions underway in the process. This heat is used to quicken production
and to increase the rate and efficiency of the ammonia that is produced.

Essential Pozzolanic Stabilization Reactions

$CaO + H_2O \longrightarrow Ca(OH)_2$
[Lime + Water ⟶ Hydrated Lime/Calcium Hydroxide]

$SiO_2 + Ca(OH)_2 + H_2O \longrightarrow CaO{\cdot}SiO_2{\cdot}nH_2O$
[Silicon Dioxide + Calcium Hydroxide ⟶ Tobermorite Gel]

$Ca(OH)_2 + H_2O + Al_2O_3 \longrightarrow CaO{\cdot}Al_2O_3{\cdot}Ca(OH)_2{\cdot}nH_2O$
[Hydrated Lime + Water + Aluminum Oxide ⟶ Calcium Aluminum Hydrates]

$CaO + Fe_2O_3 + Al_2O_3 + H_2O \longrightarrow Ca(OH)_2{\cdot}Al_2O_3{\cdot}Fe_2O_3{\cdot}nH_2O$
[Lime + Iron Oxide + Aluminum Oxide + Water ⟶ Calcium Ferro Aluminate Hydrates]

Figure 8.1 Essential reactions of thermophilic pozzolanic fixation processes.

The increased temperature, pH, alkaline agents, and silicate cause the increased production of gaseous ammonia. The free ammonia at a temperature of approximately 50 to 55°C (120 to 130°F) will assist in the inactivation of pathogens (especially parasites).

The process is based on the use of pozzolanic chemistries where temperature and ammonia content are controlled and used to convert municipal wastewater residuals into reusable products that are pathogen-free. The process requires that a closed system be operated to control the effective use of ammonia and to control odors. The entire process from feed to useable product typically takes place in fewer than 6 hours.

Process Modifications. The thermophilic process requires less sophisticated mixing agents (alkaline fly ash, lime, and sodium silicate) instead of Portland cement and sodium silicate. The high-temperature pathogen inactivation, because of both free ammonia levels and temperature, requires less ammonia. Finally, a thermophilic temperature enhances the product in releasing odorous ammonia and alkyl ammonia compounds in a controlled manner.

Product Quality. The product can be used as landfill cover (daily, interim, and final), agricultural agent (organic or lime), fill material, mined land reclamation material, or erosion control material. The product quality is related to moisture control, alkalinity, pH, permeability, and bearing strength.

This process is based on pozzolanic chemistries, and as a result possesses a reserve alkalinity that provides pathogen stability in order for the material to become acclimatized to the natural bacterial environment into which it is placed. Also, the pozzolanic chemistries provide long-term stability regarding metals leaching. In addition, the mechanical strength of the product as previously outlined also results from this pozzolanic form of stabilization chemistry.

ACID OXIDATION/DISINFECTION PROCESSES. These processes are physical–chemical processes involving a sequence of processing steps to produce a process to further reduce pathogens (PFRP) equivalent or Class A product. The process flow is shown in Figure 8.2.

Process Description. The process is capable of treating primary, blended, and waste-activated sludges and conditioned aerobically digested biosolids. The process is more economical when the feed is thickened before treatment. However, the feed should be less than 6% solids to accommodate mixing and ozone transfer. The feed is conditioned before ozonation. This conditioning may occur in a pretreatment conditioning tank or a vessel. Pretreatment conditioning involves warming the feed to a temperature of greater than 20°C (68°F) if necessary and acidification to a pH of approximately 2.5 by the addition of sulfuric acid. Temperature and pH are important to the ozonation and nitrite contact steps of the process.

In the ozonation step, the conditioned feed is exposed to an ozone-enriched gas in a pressure vessel. The ozone-enriched gas is diffused into the feed as it is recirculated in a pressure vessel. The ozonation step deodorizes, inactivates certain pathogens, and achieves a degree of oxidation indicated by a rise in the oxygen–reduction potential (ORP) of the sludge. Ozonation typically lasts for 30 to 60 minutes with ozone introduced to the vessel at a rate of approximately 30 g/min.

The ozonated biomass is discharged from the pressure vessel to a degassing tank where the dissolved gas comes out of solution at atmospheric pressure, is filtered through a charcoal filter, and vented to the atmosphere. The degassed sludge is pumped to a nitrite contact tank.

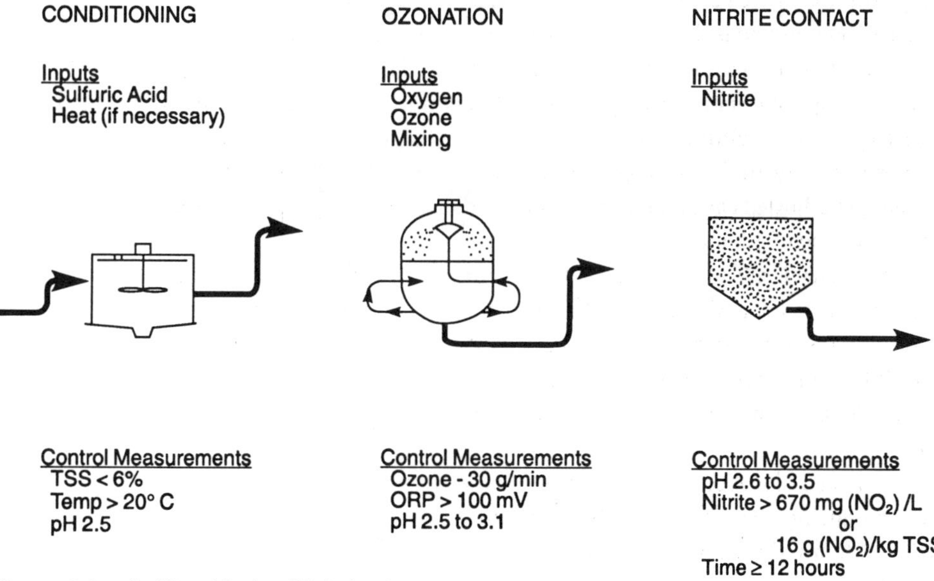

Figure 8.2 Acid oxidation/disinfection process flow.

As the ozonated sludge is pumped into the nitrite contact tank, sodium nitrite is mixed in, and the sludge is held in the tank for approximately 12 hours. The nitrite contact tank is a covered nonpressure tank that may also serve as a posttreatment equalization zone.

The required pH range for pathogen inactivation with nitrous acid is 2.6 to 3.5. The addition of sodium nitrite typically raises the pH and ORP of the sludge. However, if the pH is not greater than 2.6 after the addition of sodium nitrite, additional sodium nitrite is added until the proper pH range is achieved. The nitrite contact step reduces indicator organisms to below detection limits.

Product Quality Control. In this process, the product is disinfected and deodorized. Because volatile solids are not consumed in the process, reputrification of the treated biosolids is a possibility. Storing the biosolids at pH 3.5 or less will inhibit reputrification, the regrowth of organisms, and resulting vector attraction problems.

For PFRP or Class A treatment and associated beneficial use of the biosolids, disinfection with this process can be combined with any one of three stabilization methods to achieve long-term stability. One method is to stabilize the biosolids through a composting process. A second method is to lime-stabilize the biosolids either as a liquid or a cake by adding sufficient lime to raise the pH to greater than 12.0 for 24 hours. A third method is to dewater and dry the biosolids in drying beds with heat-drying technologies or with a nonheat pelletizing system, either of which produces a residue with a solids concentration greater than 75%.

HEAT TREATMENT/ACID DIGESTION PROCESSES. Process Description. Dewatered cake of 15 to 25% solids that has been previously screened and ground is delivered into a closed storage tank. A progressing cavity positive displacement pump continuously feeds cake from the storage tank to a plug-flow acidic vessel. Before entering the acidic vessel, the pH is adjusted to 3 with sulfuric acid, and steam is injected to raise the temperature to approximately 160°C. The process operates at 690 kPa (100 psig). The heated, pH-adjusted cake continuously enters the acidic vessel, which has a residence time of 1 hour.

Lime slurry is added to the pressurized stream exiting the acidic vessel to bring the pH to approximately 6. This pressurized stream then enters a plug-flow neutral vessel, which has a residence time of 15 minutes. The treated biosolids exit the neutral vessel through a pressure let-down system. The steam released from the pressure reduction is recycled to preheat the incoming cake.

The treated biosolids are then filtered on a recessed plate diaphragm filter press producing 60 to 70% dry solids. The filter cake is mixed with approximately 4 to 6% lime and pelletized.

Cleaning and Safety. Routine cleaning is not required by this process because there are no heat exchangers used. The sulfuric acid and lime routinely used in the process can be handled safely. To prevent injury, proper management practices and training of employees are required. Also, these facilities are designed to reduce any exposure to these chemicals. The acid addition and lime addition systems are hard-piped into the process, flanges on the pipes in these systems have flanged guards, and the acid tank is diked to contain any spill and prevent access. Access to these systems will take place only during maintenance.

Operational Considerations. The time required to assure reaction completion will vary depending on the pH. At a pH of 3 and a temperature of approximately 160°C, the material will have a residence time of approximately 1 hour in the acidic vessel.

After lime slurry is added to the acid treated material, the solids should have a residence time of 15 minutes in the neutral vessel to ensure complete reaction.

Because this process exceeds pasteurization and sterilization time requirements, and because of the inverse relationship between time and temperature in pathogen destruction, the process will assure complete pathogen destruction. This has been confirmed in tests and the process has been approved as PFRP.

The process itself has several levels of odor control:

- The solids being processed are totally enclosed in piping.
- In areas where the solids contact the atmosphere, the air is exhausted through a multistage chemical scrubber.
- The process equipment is totally enclosed in a building. Air from the building is exhausted through a scrubber.

This process uses several independent methods to assure a reduction in vector attraction, any one (or combination) of which meets the U.S. EPA Final Rule *Standards for the Use or Disposal of Sewage Sludge* regulations for vector attraction reduction (U.S. EPA, 1993):

- Lime is added to the filter cake to raise the pH to 12 or above.
- The volatile solids of the processed biosolids have been reduced by 38% from the initial material.
- If the biosolids are pelletized and dried, the solids content will exceed 75%.

HEAT/WET AIR OXIDATION PROCESS. Aqueous-phase oxidation (APO) is a destruction process that oxidizes the organic materials contained in a solids stream to primarily carbon dioxide, water, and biodegradable weak organic acids. Inorganic materials leave the process as a sterile, pathogen-free, inert solid residual that can be recycled into a number of products, including fired construction brick.

In the APO process, the solids typically undergo a 75 to 85% total solids and a 95 to 97% total volatile solids (TVS) reduction. The remaining 3 to 5% TVS content is refractory, leaving little, if any, biodegradable solids.

Process Description. Oxidizing the organic material to primarily carbon dioxide, water, and biodegradable weak organic acids is the primary process function. An added benefit is that energy liberated during the oxidation process can be recovered as hot water, steam, or electricity (Figure 8.3).

Feed at 2 to 7% solids is directed through grinders to the process oxidation vessel. The grinders ensure that the solids are no more than 3 mm in diameter to ensure good mixing and reduce the potential for plugging the downstream equipment. The feed solids can be digested or undigested. Typically, the feed is combined primary and waste-activated sludge. The raw feed can be adjusted by dilution with recycled effluent to control the process vessel influent as required to ensure proper processing.

The oxidation vessel is suspended in a deep hole, with the bottom of the vessel being approximately 1 200 to 1 500 m (4 000 to 5 000 ft) below grade. Sludge flows to the bottom of the oxidation vessel through the center pipe and then returns to the surface through the inner annulus. Oxygen is mixed in at the stoichiometric equivalent to the chemical oxygen demand (COD) of the feed. The 1 200 to 1 500 m of gas/water/sludge column results in high bottom hole pressures that allow the system to operate at the high temperatures necessary

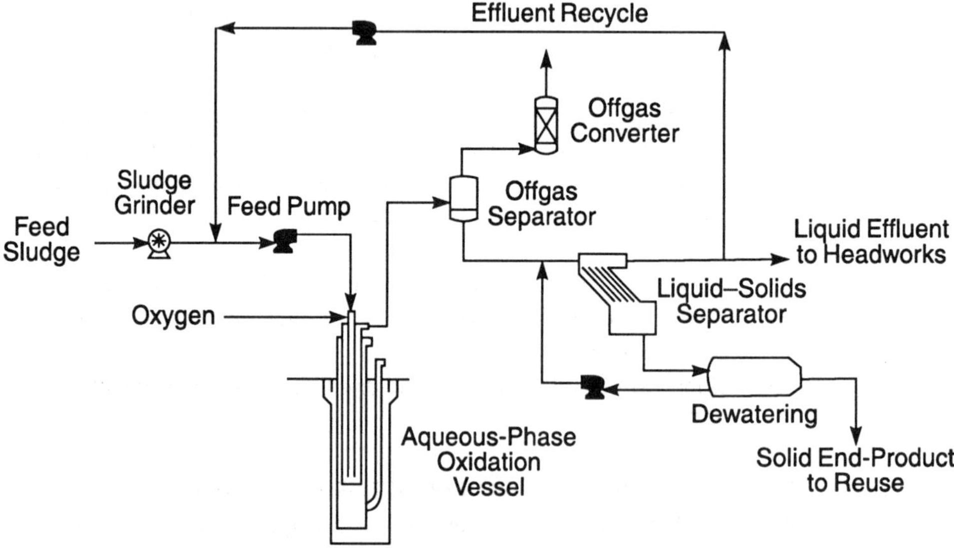

Figure 8.3   Oxidation process general flow schematic.

to attain desired treatment without boiling. The downhole vessel configuration eliminates the need for aboveground, high-pressure pumps and reactors and avoids personnel exposure to dangerous pressures and temperatures.

After processing in the oxidation vessel, the treated stream is separated into three effluent streams: gas, liquid, and solids. The separated gas stream, primarily carbon dioxide, is catalytically or thermally treated to convert any carbon monoxide to carbon dioxide. The small vent gas stream has no offensive odors.

Effluents containing sulfur, chlorine, and nitrogen products are directed to the liquid and solid phases and are not constituents of the vent gas. The liquid stream, which contains biodegradable dilute acetic acid as the primary organic product, can be returned to the headworks of the WWTP for reprocessing. The liquid stream could alternatively be treated to pretreatment or discharge standards in a small stand-alone plant.

The solid residual stream of sterile, inert materials is dewatered, resulting in 40 to 60% solids in a centrifuge or filter press without the use of flocculating chemicals. The solid stream primarily consists of inorganic carbonate, silicate, phosphate, and calcium compounds that are in the feed and do not convert to carbon dioxide and water. Typically, the volume of the solid residual stream is 4% of the incoming stream, and only 20 to 40% of the solids remain after processing. Soluble heavy metals in the stream are converted to stable solid metal oxides in the intense oxidation conditions, and are removed with the residual solids stream. The residual solid product is nonhazardous, and may be reused in a number of construction materials or can be safely landfilled.

ACTIVE SLUDGE PASTEURIZATION PROCESS. In the late 1970s, South African scientists noted a depletion of organic content in their soils. They observed that if all wastewater residuals in South Africa were to be converted to organic fertilizers, this would reduce the inorganic fertilizer demand by only 10% (thus not incurring much opposition from the inorganic fertilizer manufacturers). In the late 1980s, the active sludge pasteurization process to convert raw municipal solids to an organic fertilizer was developed from commercial fertilizer processing protocols. The process was operated at the pilot field demonstration stage in early 1990 (Bernard *et al.,* 1990).

Process Description. For optimal processing, the feed solids should be thickened to 4 to 8% to ease transport by pumping into the processing plant. If necessary, the addition of emulsification agents can refluidize the feed to facilitate pumping, processing, and spreading.

The process consists of a reactor where the feed is first mixed with anhydrous ammonia to raise the pH to 11.5. After approximately 1 hour at the high pH, all parasites, many viruses, and some bacteria are inactivated because of the high levels of free ammonia (30%). Subsequent addition of phosphoric

---

acid to the mix yields an exothermic reaction that causes the temperature to rise to approximately 65 to 70°C (150 to 160°F) within 2 minutes. This increased temperature inactivates all remaining pathogens. This same reaction causes the pH to fall from 11.5 to 7.0. The new treated organic fertilizer is exited through a vent exchanger to recapture much of the heat for the preheating of the incoming cold thickened feed. The basic scheme of the activated sludge pasteurization process is shown in Figure 8.4.

Process Modifications. The current product is a liquid fertilizer, disinfected but not stabilized. For applicability in the U.S., the product would have to be stabilized and perhaps solidified. Solidification can be achieved through dewatering and the addition of pozzolanic products such as Portland cement, silicate salts, and other pozzolanic admixtures, such as fly ash and gypsum. Stabilization can occur by adding alkaline agents that hold the pH at greater than 12, or drying to a solids content of greater than 90%.

Product Quality Control. The product quality is related to the solidification and chemical composition of the treated biosolids. The product should have the following characteristics:

- A pH greater than 12.0 and an alkalinity of 2 to 5%,
- A solids content of greater than 90%, and
- A nutrient content of greater than 2.5% nitrogen as N and greater than 4.7% phosphorus as P to be classified as a fertilizer.

Product Marketability. The product of the biosolids treatment process is a liquid organic fertilizer that can be modified to a region's specific needs. The local market for this product is viable, yet will require an outreach educational program to ensure public acceptance. The process has been operated in

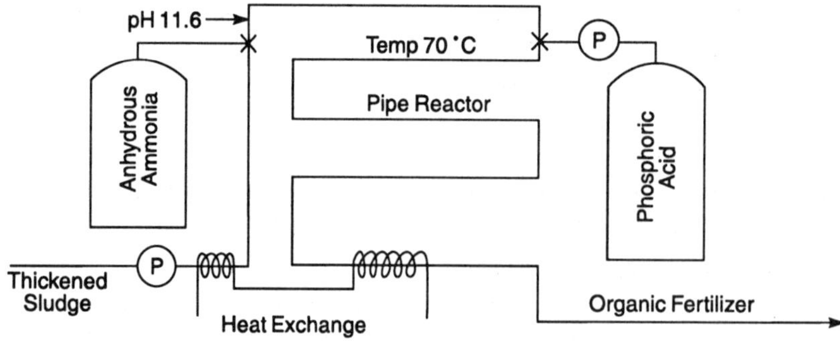

Figure 8.4   Basic scheme of the active sludge pasteurization process.

the field pilot stage, but will require development of a solidified, stabilized product for use in the U.S.

*R*EFERENCES

Bernard, J.D., *et al.* (1990) The Active Sludge Pasteurization Process. Paper presented at Int. Assoc. Water Pollut. Res. Control, Sludge Manage. Conf., Los Angeles, Calif.

U.S. Environmental Protection Agency (1993) *Standards for the Use or Disposal of Sewage Sludge. Fed. Regist.,* **58,** 32, 40 CFR Part 503.

Whittington, W.A., and Johnson, E.M. (1985) Memorandum Concerning the Application of 40 CFR Part 257 Regulations to Pathogens Reductions Preceding Land Application of Sewage Sludge or Septic Tank Pumpings. Memo to Water Div. Directors, Regions I–IX, U.S. EPA, Washington, D.C.

Appendix
Pathogen and Vector Attraction Reduction Requirements of 40 CFR Part 503

Subpart D of the U.S. EPA Part 503 Final Rule *Standards for the Use or Disposal of Sewage Sludge* regulations covers alternatives for reducing pathogens in biosolids (including domestic septage), and options for reducing the potential for biosolids to attract vectors.

The Subpart D alternatives concern the designation of biosolids as "Class A" or "Class B" in regard to pathogens. These classifications indicate the density (numbers/unit mass) of pathogens in biosolids where applicable.

If pathogens (*Salmonella* sp. bacteria, enteric viruses, and viable helminth ova) are below detectable levels, the biosolids meet the Class A designation. Biosolids are designated Class B if pathogens are detectable but have been reduced to levels that do not pose a threat to public health and the environment as long as actions are taken to prevent exposure to the biosolids after their use or disposal. When Class B biosolids are land applied, certain restrictions must be met at the application site; other requirements have to be met when Class B biosolids are surface disposed. The land application restrictions allow natural processes to further reduce pathogens in the biosolids before the public has access to the site.

CLASS A PATHOGEN REQUIREMENTS. The Part 503 rule lists six alternatives for treating biosolids so they can be classified Class A with respect to pathogens. These alternatives are summarized in Table A.1 (details are also listed in Tables A.2 and A.3). Any one of these six alternatives may be met for the biosolids to be deemed Class A.

Table A.4 lists several requirements that must be met for all six of the Class A alternatives.

CLASS B PATHOGEN REQUIREMENTS. Class B pathogen requirements can be met using one of three alternatives, as listed in Table A.5 (details are also listed in Table A.6). Unlike a Class A biosolid, in which pathogens are at levels below detectable limits, Class B biosolids may contain some pathogens. For this reason, the Class B requirements for land application of biosolids also include site restrictions that prevent crop harvesting, animal grazing, and public access for a certain period of time until environmental conditions have further reduced pathogens.

Table A.1   Alternatives for meeting Class A pathogen requirements (U.S. EPA, 1994).

In addition to meeting the requirements in one of the six alternatives listed below, the requirements in Table A.2 must be met for all six Class A alternatives.

Alternative 1: Thermally Treated Biosolids
  Biosolids must be subjected to one of four time-temperature regimes given in Table A.2.

Alternative 2: Biosolids Treated in a High pH–High Temperature Process
  Biosolids must meet specific pH, temperature, and air-drying requirements.

Alternative 3: Biosolids Treated in Other Processes
  Demonstrate that the process can reduce enteric viruses and viable helminth ova. Maintain operating conditions used in the demonstration after pathogen reduction demonstration is completed.

Alternative 4: Biosolids Treated in Unknown Processes
  Biosolids must be tested for pathogens—*Salmonella* sp. or fecal coliform bacteria, enteric viruses, and viable helminth ova—at the time the biosolids are used or disposed, or, in certain situations, prepared for use or disposal.

Alternative 5: Biosolids Treated in a PFRP
  Biosolids must be treated in one of the Processes to Further Reduce Pathogens (PFRP) (See Table A.3).

Alternative 6: Biosolids Treated in a Process Equivalent to a PFRP
  Biosolids must be treated in a process equivalent to one of the PFRPs, as determined by the permitting authority.

REQUIREMENTS FOR REDUCING VECTOR ATTRACTION. The pathogens in biosolids pose a disease risk when they are brought into contact with humans or other susceptible hosts (plant or animal). Vectors, which include flies, mosquitos, fleas, rodents, and birds, can transmit pathogens to humans and other hosts physically through contact or biologically by playing a specific role in the life cycle of the pathogen. Reducing the attractiveness of biosolids to vectors reduces the potential for transmitting diseases from pathogens in biosolids.

The Part 503 rule contains 12 options, which are summarized in Table A.7 for demonstrating reduced vector attraction for biosolids (note that Option 12 only applies to domestic septage). These requirements are designed to either reduce the attractiveness of biosolids to vectors (Options 1 through 8 and Option 12) or prevent vectors from coming in contact with the biosolids (Options 9 through 11).

Table A.2 **The four time–temperature regimes for Class A pathogen reduction under Alternative 1 (U.S. EPA, 1994).**

Regime	Applies to	Requirement	Time–temperature relationship*
A	Biosolids with 7% solids or greater (except those covered by Regime B)	Temperature of biosolids must be 50°C or higher for 20 minutes or longer	$D = \dfrac{131\ 700\ 000}{10^{0.14t}}$ (Equation 2 of Section 503.32)
B	Biosolids with 7% solids or greater in the form of small particles and heated by contact with either warmed gases or an immiscible liquid	Temperature of biosolids must be 50°C or higher for 15 seconds or longer	$D = \dfrac{131\ 700\ 000}{10^{0.14t}}$
C	Biosolids with less than 7% solids	Heated for at least 15 seconds but less than 30 minutes	$D = \dfrac{131\ 700\ 000}{10^{0.14t}}$
D	Biosolids with less than 7% solids	Temperature of biosolids is 50°C or higher with at least 30 minutes or longer contact time	$D = \dfrac{50\ 070\ 000}{10^{0.14t}}$ (Equation 3 of Section 503.32)

* D = time, days; t = temperature, °C.

Table A.3   Processes to further reduce pathogens (PFRPs) (U.S. EPA, 1994).

1. Composting

Using either the within-vessel composting method or the static aerated pile composting method, the temperature of the biosolids is maintained at 55°C or higher for 3 days.

Using the windrow composting method, the temperature of the biosolids is maintained at 55°C or higher for 15 days or longer. During the period when the compost is maintained at 55°C or higher, the windrow is turned a minimum of five times.

2. Heat drying

Biosolids are dried by direct or indirect contact with hot gases to reduce the moisture content of the biosolids to 10 percent or lower. Either the temperature of the biosolids particles exceeds 80°C or the wet bulb temperature of the gas in contact with the biosolids as the biosolids leave the dryer exceeds 80°C.

3. Heat treatment

Liquid biosolids are heated to a temperature of 180°C or higher for 30 minutes.

4. Thermophilic aerobic digestion

Liquid biosolids are agitated with air or oxygen to maintain aerobic conditions, and the mean cell residence time of the biosolids is 10 days at 55 to 60°C.

5. Beta ray irradiation

Biosolids are irradiated with beta rays from an accelerator at dosages of at least 1.0 megarad at room temperature (ca. 20°C).

6. Gamma ray irradiation

Biosolids are irradiated with gamma rays from certain isotopes, such as Cobalt 60 and Cesium 137, at room temperature (ca. 20°C).

7. Pasteurization

The temperature of the biosolids is maintained at 70°C or higher for 30 minutes or longer.

Table A.4   Pathogen requirements for all Class A Alternatives (U.S. EPA, 1994).

The following requirements must be met for *all* six Class A pathogen alternatives.

Either
- the density of fecal coliform in the biosolids must be less than 1 000 most probable numbers (MPN) per gram total solids (dry-weight basis).

or
- the density of *Salmonella* sp. bacteria in the biosolids must be less than 3 MPN per 4 grams of total solids (dry-weight basis).

Either of these requirements must be met at one of the following times:
— when the biosolids are used or disposed;
— when the biosolids are prepared for sale or give-away in a bag or other container for land application; or
— when the biosolids or derived materials are prepared to meet the requirements for exceptional quality biosolids.

Pathogen reduction must take place before or at the same time as vector attraction reduction, except when the pH adjustment, percent solids vector attraction, injection, or incorporation options are met.

Table A.5   Alternatives for meeting Class B pathogen requirements (U.S. EPA, 1994).

Alternative 1: The Monitoring of Indicator Organisms
Test for fecal coliform density as an indicator for all pathogens. The geometric mean of seven samples shall be less than 2 million MPNs per gram per total solids or less than 2 million cfus per gram of total solids at the time of use or disposal.

Alternative 2: Biosolids Treated in a PSRP
Biosolids must be treated in one of the Processes to Significantly Reduce Pathogens (PSRP) (See Table A.6).

Alternative 3: Biosolids Treated in a Process Equivalent to a PSRP
Biosolids must be treated in a process equivalent to one of the PSRPs, as determined by the permitting authority.

Table A.6 Processes to significantly reduce pathogens (PSRPs) (U.S. EPA, 1994).

1. Aerobic digestion

Biosolids are agitated with air or oxygen to maintain aerobic conditions for a specific mean cell residence time at a specific temperature. Values for the mean cell residence time and temperature shall be between 40 days at 20°C and 60 days at 15°C.

2. Air drying

Biosolids are dried on sand beds or on paved or unpaved basins. The biosolids dry for a minimum of 3 months. During 2 of the 3 months, the ambient average daily temperature is above 0°C.

3. Anaerobic digestion

Biosolids are treated in the absence of air for a specific mean cell residence time at a specific temperature. Values for the mean cell residence time and temperature shall be between 15 days at 35 to 55°C and 60 days at 20°C.

4. Composting

Using either the within-vessel, static aerated pile, or windrow composting methods, the temperature of the biosolids is raised to 40°C or higher and maintained for 5 days. For 4 hours during the 5-day period, the temperature in the compost pile exceeds 55°C.

5. Lime stabilization

Sufficient lime is added to the biosolids to raise the pH of the biosolids to 12 after 2 hours of contact.

Table A.7   Options for meeting vector attraction reduction (U.S. EPA, 1994).

Option 1:	Meet 38 percent reduction in volatile solids content.
Option 2:	Demonstrate vector attraction reduction with additional anaerobic digestion in a bench-scale unit.
Option 3:	Demonstrate vector attraction reduction with additional aerobic digestion in a bench-scale unit.
Option 4:	Meet a specific oxygen uptake rate for aerobically digested biosolids.
Option 5:	Use aerobic processes at greater than 40°C for 14 days or longer.
Option 6:	Alkali addition under specified conditions.
Option 7:	Dry biosolids with no unstabilized solids to at least 75 percent solids.
Option 8:	Dry biosolids with unstabilized solids to at least 90 percent solids.
Option 9:	Inject biosolids beneath the soil surface.
Option 10:	Incorporate biosolids into the soil within 6 hours of application to or placement on the land.
Option 11:	Cover biosolids placed on a surface disposal site with soil or other material at the end of each operating day (note: only for surface disposal).
Option 12:	Alkaline treatment of domestic septage to pH 12 or above for 30 minutes without adding more alkaline material.

REFERENCE

U.S. Envirronmental Protection Agency (1994) *A Plain English Guide to the EPA Part 503 Biosolids Rule.* EPA-832/R-93-003, Office Wastewater Manage., Washington, D.C.

Index

A

autothermal thermophilic aerobic
digestion, 39
composting, 140
diffused air mixing, 177
diffused air systems, 26
digester covers, 73
digester gas, 57, 67, 79
heat value, 68
direct dryers, 203
direct–indirect dryers, 211
disc dryers, 208
dissolved oxygen, aerobic
digestion, 19
distribution/marketing
alkaline stabilization, 194
thermal drying, 214
Downes floating cover, 73
dry lime stabilization, 150
dryers
conduction, 207
construction material, 212
direct, 203
direct–indirect, 211
disc, 208
flash, 203
fluid bed, 207
hollow-flight, 208
indirect, 207
infrared, 211
multiple-effect evaporation, 210
paddle, 208
pneumatic conveyor, 203
rotary, 205
selection, 211
drying
alkaline stabilization, 188
compost, 112
thermal, 201
dust control, alkaline stabilization,
187

E

earth filters, odor control, 40
economics
alkaline stabilization, 182

thermal drying, 216
egg-shaped digester, 73
emerging processes, 217
emissions control, thermal drying,
213
energy balance, composting, 108
energy production, autothermal
thermophilic aerobic digestion,
34
energy requirements, alkaline
stabilization, 168
energy use, anaerobic digestion, 70
explosion hazards, thermal drying,
213
external heat exchangers, 53

F

F:M, autothermal thermophilic
aerobic digestion feed, 36
facility layout, alkaline
stabilization, 180
facility siting, odor control, 138
feed characteristics
aerobic digestion, 16
alkaline stabilization, 157
anaerobic digestion, 54
autothermal thermophilic aerobic
digestion, 36
feed equipment
alkaline stabilization, 173
thermal drying, 212
filamentous bacteria, 28
filters, earth, 140
fire hazards, thermal drying, 213
fixation processes, 219
flash dryers, 203
floating covers, digester, 73
fluid bed dryers, 207
foaming
aerobic digestion, 28
anaerobic digestion, 69
autothermal thermophilic aerobic
digestion, 39

G

gas handling, anaerobic digestion, 79

gas holder covers, 73

gas injection mixing, anaerobic digestion, 81

gas injection system, anaerobic digestion, 69

gas production, anaerobic digestion, 56, 58, 67

gas scrubbing, 42

grit control, anaerobic digestion, 69

H

heat exchangers, anaerobic digestion, 84

heat recovery
autothermal thermophilic aerobic digestion, 41
thermal drying, 214

heat-transfer coefficient, 71, 86

heat transfer, thermal drying, 201

heat treatment/acid digestion processes, 222

heat/wet air oxidation processes, 223

heating, anaerobic digestion, 52, 53, 70, 84

heavy metals, anaerobic digestion, 90

hollow-flight dryers, 208

horizontal plug-flow reactors, 128

hydraulic retention time, anaerobic digestion, 62

I

indirect dryers, 202, 207

infrared dryers, 211

inhibition, anaerobic digestion, 89

intermediate storage, alkaline stabilization, 169

internal heat exchangers, 53

in-vessel composting, 107, 122

L

labor, see staffing

lime
dosing, 159
handling, 175
scaling problems, 176
slaking, 175
storage, 173

liquid lime stabilization, 149

low-temperature operation, 30

M

maintenance
alkaline stabilization, 185
anaerobic digestion, 87
autothermal thermophilic aerobic digestion, 43
heat treatment/acid digestion, 222

marketing
active sludge pasteurization, 226
alkaline stabilization, 194
compost, 142
thermal drying, 214

mass balance, 2
alkaline stabilization, 170
composting, 114

materials handling
alkaline stabilization, 166
composting, 123

maturation, composting, 111

mechanical mixing, 177
anaerobic digestion, 81

membrane covers, 74

mesophilic digestion, 52

methane production, 59, 97

microbiology
anaerobic digestion 50

process performance, alkaline
stabilization, 188
procurement, alkaline stabilization,
191
product marketing, thermal drying,
214
product quality
acid oxidation/disinfection, 222
active sludge pasteurization, 226
alkaline stabilization, 172
pozzolanic fixation processes,
220
pumps, anaerobic digestion, 78

R

radiation drying, 202
rag control, anaerobic digestion, 69
reaction rate constant, aerobic
digestion, 20
reactors
autothermal thermophilic aerobic
digestion, 40
composting, 124
recordkeeping, composting, 141
respiration rate, composting, 121
rotary dryers, 205

S

safety
alkaline stabilization, 184
anaerobic digestion, 87
composting, 142
heat treatment/acid digestion, 222
thermal drying, 214
scaling
anaerobic digestion, 96
lime, 175
screening
composting, 141
odors, 139
scrubbers, odor control, 140

scum control, anaerobic digestion,
69
sidestreams
alkaline stabilization, 188
anaerobic digestion, 93
thermal drying, 214
single-stage digestion, 72
site layout
alkaline stabilization, 181
thermal drying, 213
site preparation, alkaline
stabilization, 180
sludge volumes, anaerobic
digestion, 61
solids balance, 2
solids handling, alkaline
stabilization, 173
solids reduction, anaerobic
digestion, 56
solids retention time
aerobic digestion, 21
anaerobic digestion, 55, 62
autothermal thermophilic aerobic
digestion, 36
specific oxygen uptake rate, aerobic
digestion, 20, 21
stabilization
dry lime, 150
liquid lime, 149
process comparison, 7
staffing
alkaline stabilization, 183
thermal drying, 214
start-up, alkaline stabilization, 183
static pile composting, 106, 121
static piles, odors, 139
steam injection, anaerobic
digestion, 53
stilling wells, aerobic digestion, 25
storage
alkaline materials, 173
alkaline stabilization, 169
composting, 125
thermal drying, 213
struvite, anaerobic digestion, 96
supernatant
anaerobic digestion, 94

quality, 27
surface aerators, 26
survey, composting, 130, 144

T

tanks
  aerobic digestion, 19, 22, 25
  anaerobic digestion, 61, 73
  autothermal thermophilic aerobic
    digestion, 35, 40
  mixing, 176
temperature
  aerobic digestion, 17
  alkaline stabilization, 159
  anaerobic digestion, 55
  autothermal thermophilic aerobic
    digestion, 34
  composting, 118, 141
thermal drying
  backmixing, 212
  constant-rate, 201
  direct dryers, 203
  dryer selection, 211
  dryers, 203
  economics, 214
  emissions control, 213
  explosion hazards, 213
  falling-rate, 201
  feed equipment, 212
  feed moisture, 212
  heat recovery, 214
  heat transfer, 201
  odor control, 213
  pilot testing, 216
  process description, 203
  product marketing, 214
  safety, 214
  sidestreams, 214
  site layout, 213
  staffing, 214
  stages, 200
  storage, 213
  warm-up, 201
thermophilic aerobic digestion, 29
thermophilic digestion, 53, 100

thickening, anaerobic digestion, 70
toxic effects, anaerobic digestion, 90
turbine aerators, 27
turnover time, digester, 69
two-stage digestion, 54, 72, 99

V

vector reduction
  aerobic digestion, 20, 21
  autothermal thermophilic aerobic
    digestion, 43
  heat treatment/acid digestion, 223
vertical plug-flow reactors, 125
virus reduction, alkaline
    stabilization, 191
volatile solids reduction
  aerobic digestion, 16, 20
  anaerobic digestion, 55, 66
  autothermal thermophilic aerobic
    digestion, 41
volume, anaerobic digester, 65

W

wet air oxidation processes, 223
Wiggins floating cover, 73
windrow composting, 106, 122